ASE Test Preparation

Automotive Technician Certification Series

Advanced Engine Performance (L1)
5th Edition

DELMAR
CENGAGE Learning

Australia • Brazil • Japan • Korea • Mexico • Singapore • Spain • United Kingdom • United States

ASE Test Preparation: Automotive Technician Certification Series, Advanced Engine Performance (L1), Fifth Edition

Vice President, Technology and Trades Professional Business Unit: Gregory L. Clayton

Director, Professional Transportation Industry Training Solutions: Kristen L. Davis

Product Manager: Katie McGuire

Editorial Assistant: Danielle Filippone

Director of Marketing: Beth A. Lutz

Marketing Manager: Jennifer Barbic

Senior Production Director: Wendy Troeger

Production Manager: Sherondra Thedford

Content Project Management: PreMediaGlobal

Senior Art Director: Benjamin Gleeksman

Section Opener Image: © Christian Delbert/ www.shutterstock.com

For product information and technology assistance, contact us at
Cengage Learning Customer & Sales Support, 1-800-354-9706

For permission to use material from this text or product, submit all requests online at **www.cengage.com/permissions**.
Further permissions questions can be e-mailed to
permissionrequest@cengage.com

ISBN-13: 978-1-111-12713-8

ISBN-10: 1-111-12713-1

Delmar
Executive Woods
5 Maxwell Drive
Clifton Park, NY 12065
USA

Cengage Learning is a leading provider of customized learning solutions with office locations around the globe, including Singapore, the United Kingdom, Australia, Mexico, Brazil, and Japan. Locate your local office at **www.cengage.com/global**

Cengage Learning products are represented in Canada by Nelson Education, Ltd.

To learn more about Delmar, visit **www.cengage.com/delmar**

Purchase any of our products at your local bookstore or at our preferred online store **www.cengagebrain.com**

Notice to the Reader

Publisher does not warrant or guarantee any of the products described herein or perform any independent analysis in connection with any of the product information contained herein. Publisher does not assume, and expressly disclaims, any obligation to obtain and include information other than that provided to it by the manufacturer. The reader is expressly warned to consider and adopt all safety precautions that might be indicated by the activities described herein and to avoid all potential hazards. By following the instructions contained herein, the reader willingly assumes all risks in connection with such instructions. The publisher makes no representations or warranties of any kind, including but not limited to, the warranties of fitness for particular purpose or merchantability, nor are any such representations implied with respect to the material set forth herein, and the publisher takes no responsibility with respect to such material. The publisher shall not be liable for any special, consequential, or exemplary damages resulting, in whole or part, from the readers' use of, or reliance upon, this material.

Printed in the United States of America
3 4 5 6 7   23 22 21 20 19

Table of Contents

Delmar, a part of Cengage Learning, is very pleased that you have chosen to use our ASE Test Preparation Guide to help prepare yourself for the Advanced Engine Performance (L1) ASE certification examination. This guide is designed to help prepare you for your actual exam by providing you with an overview and introduction of the testing process, introducing you to the task list for the Advanced Engine Performance (L1) certification exam, giving you an understanding of what knowledge and skills you are expected to have in order to successfully perform the duties associated with each task area, and providing you with several preparation exams designed to emulate the live exam content in hopes of assessing your overall exam readiness.

If you have a basic working knowledge of the discipline you are testing for, you will find this book is an excellent guide, helping you understand the "must know" items needed to successfully pass the ASE certification exam. This manual is not a textbook. Its objective is to prepare the individual who has the existing requisite experience and knowledge to attempt the challenge of the ASE certification process. This guide cannot replace the hands-on experience and theoretical knowledge required by ASE to master the vehicle repair technology associated with this exam. If you are unable to understand more than a few of the preparation questions and their corresponding explanations in this book, it could be that you require either more shop-floor experience or further study.

This book begins by providing an overview of, and introduction to, the testing process. This section outlines what we recommend you do to prepare, what to expect on the actual test day, and overall methodologies for your success. This section is followed by a detailed overview of the ASE task list to include explanations of the knowledge and skills you must possess to successfully answer questions related to each particular task. After the task list, we provide six sample preparation exams for you to use as a means of evaluating areas of understanding, as well as areas requiring improvement in order to successfully pass the ASE exam. Delmar is the first and only test preparation organization to provide so many unique preparation exams. We enhanced our guides to include this support as a means of providing you with the best preparation product available. Section 6 of this guide includes the answer keys for each preparation exam, along with the answer explanations for each question. Each answer explanation also contains a reference back to the related task or tasks that it assesses. This will provide you with a quick and easy method for referring back to the task list whenever needed. The last section of this book contains blank answer sheet forms you can use as you attempt each preparation exam, along with a glossary of terms.

OUR COMMITMENT TO EXCELLENCE

Thank you for choosing Delmar, Cengage Learning for your ASE test preparation needs. All of the writers, editors, and Delmar staff have worked very hard to make this test preparation guide second to none. We feel confident that you will find this guide easy to use and extremely beneficial as you prepare for your actual ASE exam.

Delmar, Cengage Learning has sought out the best subject-matter experts in the country to help with the development of *ASE Test Preparation: Automotive Technician Certification Series,*

Advanced Engine Performance (L1), 5th Edition. Preparation questions are authored and then reviewed by a group of certified, subject-matter experts to ensure the highest level of quality and validity to our product.

If you have any questions concerning this guide or any guide in this series, please visit us on the web at **http://www.trainingbay.cengage.com**.

For web-based online test preparation for ASE certifications, please visit us on the web at **http://www.techniciantestprep.com/** to learn more.

ABOUT THE SERIES ADVISOR

Mike Swaim has been an Automotive Technology Instructor at North Idaho College, Coeur d'Alene, Idaho, since 1978. He is an Automotive Service Excellence (ASE) Certified Master Technician since 1974 and holds a Lifetime Certification from Mobile Air Conditioning Society. Swaim served as Series Advisor to all nine of the 2011 Automotive Technician/Light Truck Technician Certification Tests (A Series) of Delmar, Cengage Learning ASE Test Preparation titles, and is the author of *ASE Test Preparation: Automotive Technician Certification Series, Undercar Specialist Designation (X1), 5th Edition.*

ASE began as the National Institute for Automotive Service Excellence (NIASE). It was founded as a non-profit, independent entity in 1972 by a group of industry leaders with the single goal of providing a means for consumers to distinguish between incompetent and competent technicians. It accomplishes this goal through the testing and certification of repair and service professionals. Though it is still known as the National Institute for Automotive Service Excellence, it is now called "ASE" for short.

Today, ASE offers more than 40 certification exams in automotive, medium/heavy duty truck, collision repair and refinish, school bus, transit bus, parts specialist, automobile service consultant, and other industry-related areas. At this time there are more than 385,000 professionals nationwide with current ASE certifications. These professionals are employed by new car and truck dealerships, independent repair facilities, fleets, service stations, franchised service facilities, and more.

ASE's certification exams are industry-driven and cover practically every on-highway vehicle service segment. The exams are designed to stress the knowledge of job-related skills. Certification consists of passing at least one exam and documenting two years of relevant work experience. To maintain certification, those with ASE credentials must be re-tested every five years.

While ASE certifications are a targeted means of acknowledging the skills and abilities of an individual technician, ASE also has a program designed to provide recognition for highly qualified repair, support, and parts businesses. The Blue Seal of Excellence Recognition Program allows businesses to showcase their technicians and their commitment to excellence. One of the requirements of becoming Blue Seal recognized is that the facility must have a minimum of 75 percent of their technicians ASE certified. Additional criteria apply, and program details can be found on the ASE website.

ASE recognized that educational programs serving the service and repair industry also needed a way to be recognized as having the faculty, facilities, and equipment to provide a quality education to students wanting to become service professionals. Through the combined efforts of ASE, industry, and education leaders, the non-profit organization entitled the National Automotive Technicians Education Foundation (NATEF) was created in 1983 to evaluate and recognize academic programs. Today more than 2,000 educational programs are NATEF certified.

For additional information about ASE, NATEF, or any of their programs, the following contact information can be used:

National Institute for Automotive Service Excellence (ASE)

101 Blue Seal Drive S.E.

Suite 101

Leesburg, VA 20175

Telephone: 703-669-6600

Fax: 703-669-6123

Website: **www.ase.com**

Participating in the National Institute for Automotive Service Excellence (ASE) voluntary certification program provides you with the opportunity to demonstrate you are a qualified and skilled professional technician that has the "know-how" required to successfully work on today's modern vehicles.

EXAM ADMINISTRATION

> *Note:* After November 2011, ASE will no longer offer paper and pencil certification exams. There will be no Winter testing window in 2012, and ASE will offer and support CBT testing exclusively starting in April 2012.

ASE provides computer-based testing (CBT) exams, which are administered at test centers across the nation. It is recommended that you go to the ASE website at *http://www.ase.com* and review the conditions and requirements for this type of exam. There is also an exam demonstration page that allows you to personally experience how this type of exam operates before you register.

CBT exams are available four times annually, for two-month windows, with a month of no testing in between each testing window:

- January/February—Winter testing window
- April/May—Spring testing window
- July/August—Summer testing window
- October/November—Fall testing window

Please note, testing windows and timing may change. It is recommended you go to the ASE website at *http://www.ase.com* and review the latest testing schedules.

UNDERSTANDING TEST QUESTION BASICS

ASE exam questions are written by service industry experts. Each question on an exam is created during an ASE-hosted "item-writing" workshop. During these workshops, expert service representatives from manufacturers (domestic and import), aftermarket parts and equipment manufacturers, working technicians, and technical educators gather to share ideas and convert them into actual exam questions. Each exam question written by these experts must then survive review by all members of the group. The questions are designed to address the practical application of repair and diagnosis knowledge and skills practiced by technicians in their day-to-day work.

After the item-writing workshop, all questions are pre-tested and quality-checked on a national sample of technicians. Those questions that meet ASE standards of quality and accuracy are included in the scored sections of the exams; the "rejects" are sent back to the drawing board or discarded altogether.

Depending on the topic of the certification exam, you will be asked between 40 and 80 multiple-choice questions. You can determine the approximate number of questions you can expect to be asked during the Advanced Engine Performance (L1) certification exam by reviewing the task list in Section 4 of this book. The five-year recertification exam will cover this same content; however, the number of questions for each content area of the recertification exam will be reduced by approximately one-half.

> *Note:* Exams may contain questions that are included for statistical research purposes only. Your answers to these questions will not affect your score, but since you do not know which ones they are, you should answer all questions in the exam.

Using multiple criteria, including cross-sections by age, race, and other background information, ASE is able to guarantee that exam questions do not include bias for or against any particular group. A question that shows bias toward any particular group is discarded.

TEST-TAKING STRATEGIES

Before beginning your exam, quickly look over the exam to determine the total number of questions that you will need to answer. Having this knowledge will help you manage your time throughout the exam to ensure you have enough available to answer all of the questions presented. Read through each question completely before marking your answer. Answer the questions in the order they appear on the exam. Leave the questions blank that you are not sure of and move on to the next question. You can return to those unanswered questions after you have finished the others. These questions may actually be easier to answer at a later time, once your mind has had additional time to consider them on a subconscious level. In addition, you might find information in other questions that will help you recall the answers to some of them.

Multiple-choice exams are sometimes challenging because there are often several choices that may seem possible, or partially correct, and therefore it may be difficult to decide on the most appropriate answer choice. The best strategy, in this case, is to first determine the correct answer before looking at the answer options. If you see the answer you decided on, you should still be careful to examine the other answer options to make sure that none seems more correct than yours. If you do not know or are not sure of the answer, read each option very carefully and try to eliminate those options that you know are incorrect. That way, you can often arrive at the correct choice through a process of elimination.

If you have gone through the entire exam, and you still do not know the answer to some of the questions, *then guess.* Yes, guess. You then have at least a 25 percent chance of being correct. While your score is based on the number of questions answered correctly, any question left blank, or unanswered, is automatically scored as incorrect.

There is a lot of "folk" wisdom on the subject of test taking that you may hear about as you prepare for your ASE exam. For example, there are those who would advise you to avoid response options that use certain words such as *all, none, always, never, must,* and *only,* to name a few. This, they claim, is because nothing in life is exclusive. They would advise you to choose response options that use words that allow for some exception, such as *sometimes, frequently, rarely, often, usually, seldom,* and *normally.* They would also advise you to avoid the first and last option (A or D) because exam writers, they feel, are more comfortable if they put the correct answer in the middle (B or C) of the choices.

Another recommendation often offered is to select the option that is either shorter or longer than the other three choices because it is more likely to be correct. Some would advise you to never change an answer since your first intuition is usually correct. Another area of "folk" wisdom focuses specifically on any repetitive patterns created by your question responses (e.g., A, B, C, A, B, C, A, B, C).

Many individuals may say that there are actual grains of truth in this "folk" wisdom, and whereas with some exams, this may prove true, it is not relevant in regard to the ASE certification exams. ASE validates all exam questions and test forms through a national sample of technicians, and only those questions and test forms that meet ASE standards of quality and accuracy are included in the scored sections of the exams. Any biased questions or patterns are discarded altogether, and therefore, it is highly unlikely you will experience any of this "folk" wisdom on an actual ASE exam.

PREPARING FOR THE EXAM

Delmar, Cengage Learning wants to make sure we are providing you with the most thorough preparation guide possible. To demonstrate this, we have included hundreds of preparation questions in this guide. These questions are designed to provide as many opportunities as possible to prepare you to successfully pass your ASE exam. The preparation approach we recommend and outline in this book is designed to help you build confidence in demonstrating what task area content you already know well while also outlining what areas you should review in more detail prior to the actual exam.

We recommend that your first step in the preparation process should be to thoroughly review Section 3 of this book. This section contains a description and explanation of the type of questions you'll find on an ASE exam.

Once you understand how the questions will be presented, we then recommend that you thoroughly review Section 4 of this book. This section contains information that will help you establish an understanding of what the exam will be evaluating, and specifically, how many questions to expect in each specific task area.

As your third preparatory step, we recommend you complete your first preparation exam, located in Section 5 of this book. Answer one question at a time. After you answer each question, review the answer and question explanation information, located in Section 6. This section will provide you with instant response feedback, allowing you to gauge your progress, one question at a time, throughout this first preparation exam. If after reading the question explanation you do not feel you understand the reasoning for the correct answer, go back and review the task list overview (Section 4) for the task that is related to that question. Included with each question explanation is a clear identifier of the task area that is being assessed (e.g., Task A.1). If at that point you still do not feel you have a solid understanding of the material, identify a good source of information on the topic, such as an educational course, textbook or other related source of topical learning, and do some additional studying.

After you have completed your first preparation exam and have reviewed your answers, you are ready to complete your next preparation exam. A total of six practice exams are available in Section 5 of this book. For your second preparation exam, we recommend that you answer the questions as if you were taking the actual exam. Do not use any reference material or allow any interruptions in order to get a feel for how you will do on the actual exam. Once you have answered all of the questions, grade your results using the Answer Key in Section 6. For every question that you gave an incorrect answer to, study the explanations to the answers and/or the overview of the related task areas. Try to determine the root cause for missing the question. The easiest thing to correct is learning the correct technical content. The hardest thing to correct are behaviors that lead you to an incorrect conclusion. If you knew the information but still got the question incorrect, there is likely a test-taking behavior that will need to be corrected. An example of this would be reading too quickly and skipping over words that affect your reasoning. If you can identify what you did that caused you to answer the question incorrectly, you can eliminate that cause and improve your score.

Here are some basic guidelines to follow while preparing for the exam:

- Focus your studies on those areas you are weak in.
- Be honest with yourself when determining if you understand something.
- Study often but for short periods of time.
- Remove yourself from all distractions when studying.
- Keep in mind that the goal of studying is not just to pass the exam; the real goal is to learn.
- Prepare physically by getting a good night's rest before the exam, and eat meals that provide energy but do not cause discomfort.
- Arrive early to the exam site to avoid long waits as test candidates check in.
- Use all of the time available for your exams. If you finish early, spend the remaining time reviewing your answers.
- Do not leave any questions unanswered. If absolutely necessary, guess. All unanswered questions are automatically scored as incorrect.

Here are some items you will need to bring with you to the exam site:

- A valid government or school-issued photo ID
- Your test center admissions ticket
- A watch (not all test sites have clocks)

Note: Books, calculators, and other reference materials are not allowed in the exam room. The exceptions to this list are English-Foreign dictionaries, or glossaries. All items will be inspected before and after testing. You will be provided with a new, unmarked copy of the Composite Vehicle Manual. The regular L1 certification exam will ask approximately 10 questions based on a "generic" vehicle. The Composite Vehicle Manual provides an in-depth description of this "generic" vehicle, along with specifications and wiring diagrams that must be referred to when answering questions that involve the composite vehicle. It is advised that you study the Composite Vehicle Manual before taking the L1 ASE exam.

WHAT TO EXPECT DURING THE EXAM

When taking a CBT exam, as soon as you are seated in the testing center, you will be given a brief tutorial to acquaint you with the computer-delivered test prior to taking your certification exam(s). The CBT exams allow you to select only one answer per question. You can also change your answers as many times as you like. When you select a second answer choice, the CBT will automatically unselect your first answer choice. If you want to skip a question to return to later, you can utilize the "flag" feature, which will allow you to quickly identify and review questions whenever you are ready. Prior to completing your exam, you will also be provided with an opportunity to review your answers and address any unanswered questions.

TESTING TIME

CBT Exams

Each individual ASE CBT exam has a fixed time limit. Individual exam times will vary based upon exam area and will range anywhere from a half hour to two hours. You will also be given an additional 30 minutes beyond what is allotted to complete your exams to ensure you have adequate time to perform all necessary check-in procedures, complete a brief CBT tutorial, and potentially complete a post-test survey.

You can register for and take multiple CBT exams during one testing appointment. The maximum time allotment for a CBT appointment is four and a half hours. If you happen to register for so many exams that you will require more time than this, your exams will be scheduled into multiple appointments. This could mean that you have testing on both the morning and afternoon of the same day, or they could be scheduled on different days, depending on your personal preference and the test center's schedule.

It is important to understand that if you arrive late for your CBT test appointment, you will not be able to make up any missed time. You will only have the scheduled amount of time remaining in your appointment to complete your exam(s).

Also, while most people finish their CBT exams within the time allowed, others might feel rushed or not be able to finish the test, due to the implied stress of a specific, individual time limit allotment. Before you register for the CBT exams, you should review the number of exam questions that will be asked along with the amount of time allotted for that exam to determine whether you feel comfortable with the designated time limitation or not.

As an overall time management recommendation, you should monitor your progress and set a time limit you will follow with regard to how much time you will spend on each individual exam question. This should be based on the total number of questions you will be answering.

Also, it is very important to note that if for any reason you wish to leave the testing room during an exam, you must first ask permission. If you happen to finish your exam(s) early and wish to leave the testing site before your designated session appointment is completed, you are permitted to do so only during specified dismissal periods.

UNDERSTANDING HOW YOUR EXAM IS SCORED

You can gain a better perspective about the ASE certification exams if you understand how they are scored. ASE exams are scored by an independent organization having no vested interest in ASE or in the automotive industry. With CBT exams, you will receive your exam scores immediately.

Each question carries the same weight as any other question. For example, if there are 50 questions, each is worth 2 percent of the total score.

Your exam results can tell you:

- Where your knowledge equals or exceeds that needed for competent performance, or
- Where you might need more preparation.

Your ASE exam score report is divided into content "task" areas; it will show the number of questions in each content area and how many of your answers were correct. These numbers provide information about your performance in each area of the exam. However, because there may be a different number of questions in each content area of the exam, a high percentage of correct answers in an area with few questions may not offset a low percentage in an area with many questions.

It should be noted that one does not "fail" an ASE exam. The technician who does not pass is simply told "More Preparation Needed." Though large differences in percentages may indicate problem areas, it is important to consider how many questions were asked in each area. Since each exam evaluates all phases of the work involved in a service specialty, you should be prepared in each area. A low score in one area could keep you from passing an entire exam. If you do not pass the exam, you may take it again at any time it is scheduled to be administered.

There is no such thing as average. You cannot determine your overall exam score by adding the percentages given for each task area and dividing by the number of areas. It doesn't work that way because there generally are not the same number of questions in each task area. A task area with 20 questions, for example, counts more toward your total score than a task area with 10 questions.

Your exam report should give you a good picture of your results and a better understanding of your strengths and areas needing improvement for each task area.

Types of Questions on an ASE Exam

Understanding not only what content areas will be assessed during your exam, but how you can expect exam questions to be presented will enable you to gain the confidence you need to successfully pass an ASE certification exam. The following examples will help you recognize the types of question styles used in ASE exams and assist you in avoiding common errors when answering them.

Most initial certification tests are made up of between 40 to 80 multiple-choice questions. The five-year recertification exams will cover the same content as the initial exam; however, the actual number of questions for each content area will be reduced by approximately one-half. Refer to Section 4 of this book for specific details regarding the number of questions to expect during the initial Advanced Engine Performance (L1) certification exam.

Multiple-choice questions are an efficient way to test knowledge. To correctly answer them, you must consider each answer choice as a possibility, and then choose the answer choice that *best* addresses the question. To do this, read each word of the question carefully. Do not assume you know what the question is asking until you have finished reading the entire question.

About 10 percent of the questions on an actual ASE exam will reference an illustration. These drawings contain the information needed to correctly answer the question. The illustration should be studied carefully before attempting to answer the question. When the illustration is showing a system in detail, look over the system and try to figure out how the system works before you look at the question and the possible answers. This approach will ensure that you do not answer the question based upon false assumptions or partial data, but instead have reviewed the entire scenario being presented.

MULTIPLE-CHOICE/DIRECT QUESTIONS

The most common type of question used on an ASE exam is the direct multiple-choice style question. This type of question contains an introductory statement, called a stem, followed by four options: three incorrect answers, called distracters, and one correct answer, the key.

When the questions are written, the point is to make the distracters plausible to draw an inexperienced technician to inadvertently select one of them. This type of question gives a clear indication of the technician's knowledge.

Here is an example of a direct style question

TASK D.12

1. Which of the following would be the quickest way to test the function of the injector electrical circuit?

   A. A fuel pressure gauge

   B. A noid light

   C. A DMM

   D. A scan tool

Answer A is incorrect. A fuel pressure gauge will not check the electrical circuit of the fuel injector.

Answer B is correct. A noid light will generally be the easiest and quickest way to check the function of the fuel injector circuit.

Answer C is incorrect. While a DMM can be used to test the condition of the injector wiring, it will not easily show whether or not the circuit is being switched on and off.

Answer D is incorrect. While a scan tool may provide information about the functionality of an injector circuit, it will not provide many options for testing the circuit.

1. A vehicle with electronic throttle control has an accelerator pedal position (APP) diagnostic trouble code. The Technician wiggles the wiring harness while observing the APP sensor voltage with the scan tool. During this procedure, the sensor voltage changes. Which of the following is the most likely cause of the diagnostic trouble code?

    TASK B.8

    A. Faulty scan tool
    B. Faulty ECM
    C. Faulty APP sensor wiring
    D. Faulty ECM power supply

Answer A is incorrect. If the voltage value changed while the wiring harness was moving, there is no reason to believe the scan tool is faulty.

Answer B is incorrect. If the voltage value changed while the wiring harness was moving, there is no reason to believe the ECM is faulty.

Answer C is correct. If the voltage value changed while the wiring harness was moving, the most likely cause is the wiring harness.

Answer D is incorrect. If all other items associated with the ECM are normal and the voltage value changed while the wiring harness was moving, the most likely cause is the wiring harness.

TECHNICIAN A, TECHNICIAN B QUESTIONS

This type of question is usually associated with an ASE exam. It is, in fact, two true-false statements grouped together, such as: "Technician A says ..." and "Technician B says ...", followed by "Who is correct?"

In this type of question, you must determine whether either, both, or neither of the statements are correct. To answer this type of question correctly, you must carefully read each technician's statement and judge it on its own merit.

Sometimes this type of question begins with a statement about some analysis or repair procedure. This statement provides the setup or background information required to understand the conditions about which Technician A and Technician B are talking, followed by two statements about the cause of the concern, proper inspection, identification, or repair choices.

Analyzing this type of question is a little easier than the other types because there are only two ideas to consider, although there are still four choices for an answer.

Again, Technician A, Technician B questions are really double true-or-false questions. The best way to analyze this type of question is to consider each technician's statement separately. Ask yourself, "Is A true or false? Is B true or false?" Once you have completed an individual evaluation of each statement, you will have successfully determined the correct answer choice for the question, "Who is correct?"

An important point to remember is that an ASE Technician A, Technician B question will never have Technician A and B directly disagreeing with each other. That is why you must evaluate each statement independently.

An example of a Technician A/Technician B style question looks like this:

TASK D.5

1. Technician A says that positive fuel trim percentages indicate a rich air/fuel mixture. Technician B says that positive fuel trim percentages could be caused by vacuum leak. Who is correct?

   A.   A only

   B.   B only

   C.   Both A and B

   D.   Neither A nor B

Answer A is incorrect. Positive fuel trim percentages indicate that the ECM is adding fuel to correct for a lean mixture.

Answer B is correct. Technician B is correct. Positive fuel trim percentages indicate that the engine is operating lean and having to add fuel to correct for the lean mixture. A vacuum leak can cause this lean condition.

Answer C is incorrect. Only Technician B is correct.

Answer D is incorrect. Only Technician B is correct.

EXCEPT QUESTIONS

Another question type used on the ASE exams contains answer choices that are all correct except for one. To help easily identify this type of question, whenever it is presented in an exam, the word "EXCEPT" will always be displayed in capital letters. Furthermore, a cautionary statement will alert you to the fact that the next question is different from the ones otherwise found in the exam. With the EXCEPT type of question, only one incorrect choice will actually be listed among the options, and that incorrect choice will be the key to the question. That is, the incorrect statement is counted as the correct answer for that question.

Be careful to read these question types slowly and thoroughly; otherwise, you may overlook what the question is actually asking and answer the question by selecting the first correct statement.

An example of this type of question would appear as follows:

TASK E.12

1. All of the following could be a result of a malfunctioning positive crankcase ventilation (PCV) system EXCEPT:

   A.   Oil leaks

   B.   MIL illumination

   C.   Unstable idle

   D.   Increased fuel economy

Answer A is incorrect. A malfunctioning PCV system could cause oil leaks.

Answer B is incorrect. A vacuum leak in the PCV system could cause the MIL to illuminate.

Answer C is incorrect. Since the PCV system provides a high percentage of the engine's idle airflow, a malfunction in the system could cause an unstable idle.

Answer D is correct. A malfunction in the PCV system will not cause increased fuel economy.

LEAST LIKELY QUESTIONS

LEAST LIKELY questions are similar to EXCEPT questions. Look for the answer choice that would be the LEAST LIKELY cause (most incorrect) of the described situation. To help easily identify these type of questions, whenever they are presented in an exam, the words "LEAST LIKELY" will always be displayed in capital letters. In addition, you will be alerted before a LEAST LIKELY question is posed. Read the entire question carefully before choosing your answer.

An example of this type of question is shown below:

1. A vehicle overheats when pulling a trailer. Which of the following would be the LEAST LIKELY cause?

   A.   Slipping fan clutch
   B.   Seized fan clutch
   C.   Restricted charge air cooler
   D.   Restricted radiator

TASK D.2

Answer A is incorrect. A slipping fan clutch may not fully engage and would fail to provide sufficient air flow across the radiator to keep the engine cool.

Answer B is correct. A seized fan clutch would run all the time; this may cause a low power complaint but would not cause the engine to overheat.

Answer C is incorrect. A restricted charge air cooler would also restrict the air flow across the radiator. This could result in an engine overheating condition.

Answer D is incorrect. A restricted radiator could result in an overheated engine.

SUMMARY

The question styles outlined in this section are the only ones you will encounter on any ASE certification exam. ASE does not use any other types of question styles, such as fill-in-the-blank, true/false, word-matching, or essay. ASE also will not require you to draw diagrams or sketches to support any of your answer selections, although any of the described question styles may include illustrations, charts, or schematics to clarify a question. If a formula or chart is required to answer a question, it will be provided for you.

INTRODUCTION

This section of the book outlines the content areas or *task list* for this specific certification exam, along with a written overview of the content covered in the exam.

The task list describes the actual knowledge and skills necessary for a technician to successfully perform the work associated with each skill area. This task list is the fundamental guideline you should use to understand what areas you can to expect to be tested on, as well as how each individual area is weighted to include the approximate number of questions you can expect to be given for that area during the ASE certification exam. It is important to note that the number of exam questions for a particular area is to be used as a guideline only. ASE advises that the questions on the exam may not equal the number listed on the task list. The task lists are specifically designed to tell you what ASE expects you to know how to do and to help you prepare to be tested.

Similar to the role this task list will play in regard to the actual ASE exam, Delmar, Cengage Learning has developed six preparation exams, located in Section 5 of this book, using this task list as a guide. It is important to note that although both ASE and Delmar, Cengage Learning use the same task list as a guideline for creating these test questions, none of the test questions you will see in this book will be found in the actual, live ASE exams. This is true for any test preparatory material you use. Real exam questions are *only* visible during the actual ASE exams.

Task List at a Glance

The Advanced Engine Performance (L1) task list focuses on six core areas, and you can expect to be asked approximately 50 questions on your certification exam, broken out as outlined here:

- A. General Powertrain Diagnosis (6 questions)
- B. Computerized Powertrain Controls Diagnosis (Including OBD-II) (16 questions)
- C. Ignition System Diagnosis (6 questions)
- D. Fuel Systems and Air Induction Systems Diagnosis (8 questions)
- E. Emission Control System Diagnosis (8 questions)
- F. I/M Failure Diagnosis (6 questions)

Based upon this information, the graph shown here is a general guideline demonstrating which areas will have the most focus on the actual certification exam. This data may help you prioritize your time when preparing for the exam.

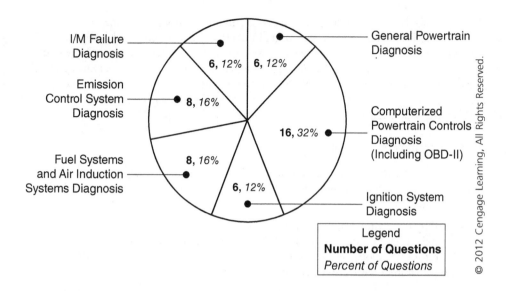

I/M Failure Diagnosis — **6**, *12%*

General Powertrain Diagnosis — **6**, *12%*

Emission Control System Diagnosis — **8**, *16%*

Computerized Powertrain Controls Diagnosis (Including OBD-II) — **16**, *32%*

Fuel Systems and Air Induction Systems Diagnosis — **8**, *16%*

Ignition System Diagnosis — **6**, *12%*

Legend
Number of Questions
Percent of Questions

> *Note:* The actual number of questions you will be given on the ASE certification exam may vary slightly from the information provided in the task list, as exams may contain questions that are included for statistical research purposes only. Don't forget that your answers to these research questions will not affect your score.

A. General Powertrain Diagnosis (6 Questions)

1. Verify customer concern; determine if the concern is the result of a malfunction or normal system operation.

The first step in any diagnosis is to verify the customer's concern. When verified, note the conditions present that caused the concern to occur, such as the amount of time the vehicle has been driven, engine temperature, engine load, RPM, vehicle speed, etc. This information will be useful when verifying the repair. If the concern cannot be duplicated and verified, it may be nearly impossible to diagnose the concern or verify that it is fixed if repairs are made. Once the concern has been verified, determine if the concern is normal system operation or an actual malfunction of the system. Knowledge of system operation is critical to determining if the concern is normal or a malfunction that requires repair. Refer to service information for a description of normal system operation or compare with another vehicle of the same type if one is available.

2. Inspect and test for missing, modified, inoperative, or tampered powertrain mechanical components.

Changing or bypassing mechanical components on a late-model engine not only violates Federal law; it can also negatively affect vehicle performance, fuel economy, and emission control. Using a component locator manual and knowledge of the system, inspect for missing, damaged, modified, or aftermarket replacement parts. Ensure that all vacuum hoses are connected correctly and are free of damage. Also be sure that wiring harnesses are connected properly and are free of corroded connections and other damage such as melted or chafed insulation, and that they are also routed and secured properly.

Mechanical components that are often tampered with include air intake hoses, fuel-pressure regulators, fuel injectors, vacuum switches and solenoids, exhaust gas recirculation (EGR) systems, air-injection systems, intake and exhaust manifolds (replaced with unapproved aftermarket units), and other exhaust components like the catalytic converter and muffler. Altering or rendering any of the above-mentioned components inoperative could change the air/fuel mixture delivery and/or requirements of the engine, as well as increase emissions or reduce fuel economy.

Aftermarket intake and exhaust manifolds can adversely affect emissions and fuel economy. Aftermarket intake manifolds must contain any provisions for exhaust gas recirculation (EGR) that were present on the original equipment manifold. Failure to provide EGR on an engine where it was original equipment can cause increased NO_x emissions and cause detonation or pinging. Similarly, aftermarket exhaust manifolds or headers must contain original equipment provisions for exhaust heat risers, air-injection nozzles, oxygen sensors, and air fuel ratio sensors. As a general rule, replacement manifolds should be selected only from the carmaker's available options or from aftermarket suppliers who have obtained approval from government agencies.

3. Locate relevant service information.

The amount of diagnostic and service information provided for new vehicles by the manufacturers grows enormously each year. Finding the right information and recognizing what is relevant for a specific job, and what is not, is an important skill for any technician.

Diagnostic information is published by the carmakers, and much of this information is available in aftermarket manuals and online by vehicle manufacturers and independent publishers. This information includes lists of diagnostic trouble codes (DTC), test specifications, and diagnostic flow charts. Component location information, vacuum and wiring diagrams, and part numbers are also essential service data.

Technical service bulletins (TSB) are an important part of every carmaker's information systems. TSBs are intended primarily for dealership service departments, but many are available through aftermarket information sources. TSBs are often used to inform dealer service technicians about manufacturer-approved changes to control system calibrations, fixes for recurrent problems, running changes made to vehicles during a model year, updated service parts, and service manual updates. TSBs are a primary source of up-to-date information. If available, they should be consulted for relevant data for most diagnostic jobs.

4. Research system operation using technical information to determine diagnostic procedure.

The need to fully understand system and component operation is essential when diagnosing or repairing today's complex vehicles. Utilizing vehicle service, computerized repair information systems or technical service bulletins (TSB), the technician should research and become familiar with the systems on the vehicle being serviced and diagnosed. With a thorough understanding of the systems present on a particular vehicle, a diagnostic approach can be undertaken and the correct test procedures applied.

Failure to identify and understand newer systems, such as variable valve timing and intake runner controls, can lead to incorrect diagnosis of engine performance problems, such as hard starting or lack of power. Deciding on the correct diagnostic path to follow is one of the most challenging problems facing a technician servicing computerized vehicles today. Technicians must often use their knowledge and intuition when researching technical information to determine and understand engine-management system strategy. Diagnostic flow charts provided by the manufacturer do provide a valuable source of specifications for testing systems or components. When testing a component or system, a technician can use the specifications found in flow charts to determine the minimum and maximum allowable values. This is known as specification-based diagnosis and can allow the technicians' experience and ingenuity to work for them. Applying a logical and consistent diagnostic procedure is the most important skill a technician can possess. This procedure must be followed consistently to prevent overlooking the basics, which is where many problems are often found. Failing to follow a set procedure or process that verifies the basics will lead to diagnostic failures.

5. Use appropriate diagnostic procedures based on available vehicle data and service information; determine if available information is adequate to proceed with effective diagnosis.

After service information has been gathered and reviewed, the technician must decide if there is enough information to proceed with accurate diagnosis and repair. Current TSB information may be essential for a car that is one or two years old, but it may not be relevant for an older model. Information on conditions that cause a certain DTC to set may be essential when troubleshooting a specific DTC, but it may not be relevant if no codes are present. A wiring diagram may be essential for complicated electrical troubleshooting and repair; for some repairs, following the wire color codes on the vehicle may be sufficient.

Probably the most important piece of service information is an accurate verification of the complaint or the problem. A diagnostic technician should get all possible information from the car owner or driver about driveability symptoms and about any recent work done on the car. If a symptom is present during the diagnosis and can be verified by the technician, the job is often easier. If the symptom is not present, getting a complete description from the driver is even more important.

Problems that are intermittent illustrate the need for improved communications between the customer and the technician. When a problem cannot be verified, the repair also cannot be verified. With a complete description of the problem, the technician can determine whether or not an effective diagnosis can be made.

6. Determine the relative importance of observed vehicle data.

In any set or combination of vehicle symptoms, some are more important than others. For example, if a malfunction indicator lamp (MIL) is lit, one or more DTCs are almost certainly present. Reading and troubleshooting those DTCs should be the highest-priority diagnostic tasks that relate to the symptoms of the car. If the symptoms also include a complaint about a rough idle or surging at cruising speeds, troubleshooting the DTCs and fixing their causes may also correct the other driveability problems.

Stored DTCs can have a greater impact on OBD-II systems due to the interaction of components and the monitors that an OBD-II computer runs. There is a hierarchy of diagnostic monitors in an OBD-II system. The monitors run in a specific order. If one monitor fails to pass its system, the other monitors will not run.

For example, a failed oxygen sensor will store a DTC for the sensor, but the computer also uses the sensor input to run the diagnostic monitor for the catalytic converter and possibly the EGR system. When the oxygen sensor code is set, the catalyst and EGR monitors are suspended, pending the correction of the oxygen sensor problem. Once the oxygen sensor is repaired, it is possible for another code to appear in a different system after the diagnostic monitor that was "blocking" the other monitors passes its testing criteria. It may be necessary to run all monitors and recheck for codes once a component is replaced on an OBD-II vehicle.

7. Differentiate between powertrain mechanical and electrical/electronic problems, including variable valve timing (VVT) systems.

Mechanical failures can mimic electronic problems and vice versa. Computerized engine controls depend on a mechanically sound engine. Many DTCs that appear in the computer system can actually be caused by mechanical wear or failure in the engine. For example, if the MIL is lit and a DTC is present for a manifold absolute pressure (MAP) sensor problem, an intake vacuum leak can cause the DTC to set just as easily as an electrical or sensor failure. This is just one example of how a mechanical problem can cause an input to the ECM to give a reading that may look incorrect to the technician even though the sensor is reporting the actual condition that it is measuring.

Variable valve timing (VVT) systems use electronic components to control mechanical devices. Many times, electrical solenoids are used to supply oil pressure to actuators. While the electrical part of the system may be fairly straightforward, the mechanical actuators and cam phasers rely on precise timing, installation, and oil viscosity. When a fault occurs, pay close attention to any DTCs that were set for the system, and understand what faults could cause the DTC to set. Many times, electrical faults will set DTCs while mechanical faults will not. When diagnosing mechanical faults in the VVT system, be sure to have a detailed understanding of normal system operation and the symptoms that can occur if the system were to fail.

8. Diagnose driveability problems and emission failures caused by cooling system problems.

A low coolant level can cause the engine coolant temperature (ECT) sensor to warm up unevenly. An uneven, or erratic, signal from the ECT sensor will affect fuel metering and the air/fuel mixture. This, in turn, can lead to driveability and fuel economy problems, as well as increased CO and HC emissions.

An engine that is running too hot may tend to knock, which can cause the knock sensor to signal the ECM to retard ignition timing. Poor performance and fuel economy can result. An overheated engine may also suffer from incomplete combustion and a resulting increase in HC emissions.

If a cooling system problem prevents the engine from reaching normal operating temperature, the engine may run rich or may never reach temperatures required for closed-loop operation. Poor fuel mileage and sluggish performance may result.

If coolant enters the exhaust system through a combustion chamber leak, it will contaminate the O_2 sensor and eventually prevent it from working.

9. Diagnose driveability problems and emission failures caused by engine mechanical problems.

The adaptive learning capability of most late-model engine control computers can adjust to changes and wear in the engine, but no system can compensate for base engine mechanical problems that affect compression or airflow through the engine. A technician must verify engine integrity before attempting to diagnose problems with the engine control system. Vacuum tests, relative compression tests, cranking and running compression tests, and cylinder leakage tests may all need to be performed to verify engine mechanical integrity. Engine vacuum testing can determine the presence of cylinder sealing and airflow problems in the engine. Relative compression tests can help isolate individual cylinder problems. Cranking and running compression tests done with a gauge will pinpoint cylinder sealing or airflow problems. Cylinder leakage testing can pinpoint the exact cause of cylinder sealing problems such as leaking intake or exhaust valves or head gasket failure.

An engine with a burned exhaust valve, resulting in low cylinder pressure, may show a continuously lean exhaust condition because of the additional oxygen in the exhaust due to incomplete combustion. This condition may be indicated by higher than normal O_2 readings on a four-gas exhaust analyzer or by a continuous lean-exhaust signal from the exhaust O_2 sensor.

A leaking or burned exhaust valve will not affect airflow, but it can cause incomplete combustion by not sealing the combustion chamber. A leaking intake valve can also reduce cylinder pressure and lead to incomplete combustion, and it can affect intake airflow.

Incomplete combustion resulting from low compression or a vacuum leak can let uncombined oxygen pass through the combustion chamber to the exhaust. This can cause continual lean-exhaust readings from the O_2 sensor (low voltage from a common zirconia O_2 sensor).

A worn camshaft lobe will reduce both lift and duration for its valve. Depending on whether the valve is an intake or exhaust valve, airflow into the cylinder or exhaust flow out will be affected. Cam lobe wear is often uneven among cylinders and results in rough running and uneven compression.

A timing belt or timing chain that has slipped will alter valve timing and generally affect all cylinders equally.

Leaking valves, broken or worn piston rings, and leaking head gaskets can all cause low compression. Depending on the amount of compression loss and the number of cylinders affected, the engine may run roughly and misfire.

10. Diagnose driveability problems and emission failures caused by problems or modifications in the transmission and final drive, or by incorrect tire size.

Transmission shift speeds that are too high or too low can reduce vehicle performance and fuel economy. An automatic transmission that shifts at the wrong speed can also increase exhaust emissions during an I/M emission test drive cycle. If transmission shift points are altered by installing a "shift kit," emissions may similarly be increased.

The engine control module (ECM) or transmission control module (TCM) receives temperature sensor signals, speed sensor signals, and gear position signals from the automatic transmission, as well as input signals from engine sensors. The ECM or TCM uses combinations of these signals to determine when to lock up the torque converter and when to command upshifts and downshifts. Replacing the ECM (or the PROM inside the ECM) is sometimes done to fix driveability problems associated with torque converter lock-up. If the change originates with the carmaker, it will not adversely affect emissions or fuel economy. However, aftermarket PROMs or ECM changes may adversely affect emissions and economy. The carmaker's TSBs are a good source of information about approved ECM modifications.

A torque converter that continuously locks and unlocks may create a symptom very similar to an engine misfire or surging condition. One good way to distinguish between an engine misfire or surge and a torque converter shudder is to drive the car until the symptom occurs and then lightly touch the brake pedal. Operating the brake switch commands the ECM or TCM to unlock the converter. If the symptom goes away, it was most likely related to torque converter lock-up and not to the engine. Be careful, as some engine misfires can be load dependant. With the torque converter clutch engaged, the engine will have a greater load. Secondary ignition breakdown will most likely cause a miss under heavy load conditions.

Installing oversize wheels and tires on a vehicle without a corresponding correction for the vehicle speed sensor (VSS) can cause inaccurate speed signals to be sent to the ECM or TCM. Generally, oversize wheels and tires cause a VSS-speed signal that is slower than true vehicle speed. Any inaccurate speed signal can affect torque converter lock-up, shift points, and general engine performance. Wheels and tires other than the original equipment sizes should be selected from the carmaker's approved optional sizes. The VSS signal can often be recalibrated for these optional tire sizes.

11. Diagnose driveability problems and emission failures caused by intake or exhaust system problems or modifications.

When exhaust system components are changed in an attempt to improve performance, catalytic converters are often relocated or sometimes removed altogether. Original-equipment exhaust systems are designed for the best combination of exhaust flow and any slight backpressure necessary for exhaust gas recirculation (EGR) operation. Changing pipe diameters—usually from smaller to larger—and relocating mufflers and converters can upset original exhaust flow and back pressure requirements. More often than not, vehicle performance suffers.

Any change in EGR operation caused by exhaust system modifications can lead to increased NO_x emissions and engine pinging, because EGR is a principal method used to control detonation.

Aftermarket exhaust systems must meet original-equipment specifications. In many cases, they must also be approved by government regulatory agencies. Aftermarket exhaust systems must also include exhaust oxygen sensors (O_2) of original-equipment specifications installed in original locations in the exhaust pipes or manifolds. Most importantly, a vehicle originally equipped with a catalytic converter may not have the converter removed.

A plugged catalytic converter or other restriction in the exhaust system will increase exhaust backpressure, which can lead to several driveability complaints. Most noticeably, the vehicle will lose power and perform sluggishly. Gas mileage may decrease, and the vehicle may fail an emissions test.

An exhaust leak upstream from the O_2 sensor may draw air into the exhaust stream and cause erroneous O_2 sensor readings. This, in turn, can lead to incorrect air/fuel mixture control by the ECM and may cause the system to drop out of closed loop into open-loop operation. Several different DTCs may occur.

12. Determine root cause of failures.

13. Determine root cause of multiple component failures.

14. Determine root cause of repeated component failure.

Tasks 12, 13, and 14 are all interrelated, and they are equally related to similar tasks listed for the ignition, fuel, emission control, and electronic systems. These tasks are examples of step-by-step logic and critical thinking applied by the technician.

It is not enough just to replace a failed component without determining the cause of the failure. For example, if an O_2 sensor is replaced and the new one fails in 5,000 to 10,000 miles, the root cause of the failure was not

identified. Perhaps a cooling system leak into a combustion chamber caused deposits to form on the O_2 sensor. Fixing the cooling system leak will fix the root cause of the O_2 sensor failure.

Several related components may fail from a single cause. An O_2 sensor and a catalytic converter could both suffer from the coolant leak described above. Depending on the severity of the leak and the length of time that it existed, it might be wise to check catalytic converter efficiency with a four-gas exhaust analyzer after replacing the second O_2 sensor.

The examples given illustrate the importance of finding the root cause of both multiple and repeated component failures. Other similar examples for specific problem-solving tasks relating to the ignition, fuel, and emission systems are provided later in this overview.

15. Verify effectiveness of repairs.

Once the concern has been diagnosed and repaired, the effectiveness of the repair needs to be verified. Recall the conditions that were present when the fault occurred and operate the vehicle as closely to those conditions as possible. Make sure the customer's concern is no longer present under these conditions. Operate the vehicle in a variety of other conditions as well to make sure the condition does not occur under those conditions. Refer to any scan tool data that read incorrectly while the malfunction occurred and confirm that it is now correct and accurate. Repairs may also be verified by running any tests that led to a diagnosis after the repair has been made. The vehicle should now pass these tests. It may also be wise for the technician to check any monitors that previously failed when the fault occurred. Verifying that these monitors pass after repairs are made will also verify the repair. The technician should also be sure that the repair has not created any new concerns for the customer; if so, these should be addressed before returning the vehicle.

B. Computerized Powertrain Controls Diagnosis—Including OBD-II (16 Questions)

1. Verify customer concern; determine if the concern is the result of a malfunction or normal system operation.

Before diagnosis can begin, the customer's concern must be verified. Knowledge of vehicle operation and on-board diagnostics will help the technician determine if the concern is normal system operation or the result of a malfunction. Refer to service information, and research the normal operation of the vehicle. Accurate descriptions are available to familiarize the technician with normal system operation. Apply this information to the customer's concern to determine if a malfunction exists. If the concern has been duplicated and determined to be a malfunction, note the conditions present that caused the concern to occur, such as the amount of time the vehicle has been driven, engine temperature, engine load, RPM, vehicle speed, etc. Identify any abnormal scan tool data readings that may help in diagnosis of the vehicle. Keep track of these conditions and readings, as they will aid in verifying the effectiveness of the repair.

2. Inspect and test for missing, modified, inoperative, or tampered computerized powertrain control components.

Changing, bypassing, or tampering with electronic engine control components violates Federal law and usually harms vehicle performance, fuel economy, and emission control. Part of troubleshooting an electronic control system is to inspect it for missing, damaged, or tampered components, just as any other vehicle system would be inspected.

Ensure that all vacuum hoses are correctly connected and free of damage. Also be sure that wiring harnesses are properly connected and free of damage and corroded connections. An unattached connector on a wiring harness may be a clue that a device has been removed from the system. At this point, a component locator manual and a wiring diagram can be essential in determining if a part is missing.

If a vacuum solenoid is not controlling vacuum as it is supposed to, inspect the vacuum hoses and solenoid ports to see if they have been tampered with and intentionally plugged. A coolant sensor may be disconnected or have a

resistor added to its signal wire to intentionally force the ECM to command a rich air/fuel ratio. This kind of tampering may be done in a misguided attempt to improve driveability, but it will harm fuel economy and emission control.

Similarly, inspect the throttle position sensor (TPS) for obvious misadjustment or modification. This kind of tampering is sometimes done to try to fix a rough or uneven idle, but it can throw the entire fuel-control program of the ECM out of range.

3. Locate relevant service information.

Gathering service information begins with understanding and verifying the driver's complaint or the vehicle symptoms. The technician should also find out if any recent service work has been done on the vehicle or if any aftermarket electronic accessories have been installed. If custom audio equipment or a cellular phone has been installed, these accessories may be the source of electromagnetic interference (EMI) that affects the control system. Additionally, a poor ground connection or power source for an accessory can cause problems with the vehicle's original electronic equipment. Learning about a vehicle's service history and condition should be the first act of getting information about that vehicle.

Service information for the powertrain control system can come from the same sources as other information about the car. Selecting the most relevant information can be the key to a fast and accurate diagnosis. Wiring diagrams may be particularly important for control system troubleshooting. Diagnostic flowcharts and "trouble trees" can be equally important.

4. Research system operation using technical information to determine diagnostic procedure.

For accurate and efficient diagnosis, a technician must understand the operating details of a specific system. For example, some control systems have "default" or "limp-in" operating modes that let the system continue to operate by using back-up values from its own memory in case of sensor failures. Other systems do not have such default operating modes. Technicians must know the characteristics of the particular systems they are testing.

As another example, some fuel-injection control systems have a "clear-flood" operating mode. In this mode, the ECM will reduce or cut off fuel flow during cranking if TPS voltage is above 80 percent of its range. Knowing whether or not a particular system has this characteristic can be an important part of diagnosing a no-start problem.

Additionally, adaptive-memory features vary from one system to another. This feature is the ability of an ECM to learn the operating characteristics of a particular vehicle and to compensate for age and engine wear. Resetting the adaptive memory, when applicable, is a key step in many service procedures.

Much of this information is provided in either printed or electronic service information. TSBs may also provide useful information required to diagnose the vehicle.

5. Use appropriate diagnostic procedures based on available vehicle data and service information; determine if available information is adequate to proceed with effective diagnosis.

Troubleshooting a powertrain control system often requires several series of diagnostic procedures. A technician may have to obtain a diagnostic flowchart and perform the test procedure only to find that further tests are required. The overall diagnostic process can be repetitive. The first round of diagnosis may not provide adequate information to confirm the root cause of the problem.

To diagnose general powertrain control system problems, it is necessary to have a thorough understanding of system design principals and operating strategies. Systems that comply with OBD-II diagnostic standards provide a minimum list of operating data parameters and have standardized DTCs. Accessing reliable repair and service information is essential in any diagnosis, including the parameters and criteria to set a DTC. Also, do not overlook reviewing any TBSs that could be related to the problem. This information can provide an excellent starting point for troubleshooting.

6. Determine current version of computerized powertrain control system software and updates; perform reprogramming procedures.

In the past, ECMs had a physical "chip" that contained the software for the operation of the vehicle. Today's vehicles no longer require this "chip" and instead have programmable software packages that can be updated using the factory scan tool or specialized aftermarket equipment. Both of these methods usually require a subscription to a service from which software updates can be downloaded. There are many methods to transfer this software from the computer to the ECM. Some methods transfer the software to the scan tool and then from the scan tool to the vehicle's ECM that is being programmed. Other methods have the ability to transfer the software from the computer directly to the vehicle's ECM. Regardless of the method, these software updates are intended to change operating parameters to correct known problems that the vehicle may exhibit. The best method of learning if a software update is available to correct a known issue is by checking technical service bulletins (TSBs). Many manufacturers generally do not recommend updating the software of the ECM unless directed by a bulletin or service procedure. It may, at times, be difficult to identify the current version of software that is programmed into the vehicle. This may or may not be listed in scan tool data.

Performing software updates is generally performed key on, engine off. As battery voltage is critical during reprogramming, be sure that all accessories and especially daytime running lights are off to prevent unnecessary battery drain. Many manufacturers do not recommend connecting a standard battery charger while programming, as voltage fluctuations can cause the programming process to fail and possibly damage the ECM. Specialty chargers have been developed to provide a constant DC voltage with little or no leftover AC voltage from the wall outlet. Be sure that the battery is fully charged before performing any reprogramming procedure. A full battery and charging system test may be necessary to ensure that battery voltage will remain at least 12 volts during the programming process. Refer to and follow all instructions provided by the manufacturer while programming the ECM.

7. Research OBD-II system operation to determine the enable criteria for setting and clearing diagnostic trouble codes (DTCs) and malfunction indicator lamp (MIL) operation.

Generally, when a fault has occurred that causes emissions levels to increase 1.5 times the federal limit, a DTC will set and the MIL will illuminate. Certain conditions, such as a catalyst damaging fault (a severe misfire, for example), may cause the MIL to flash. To test the operation of the MIL, turn the key on and confirm that the MIL illuminates for approximately two seconds. This verifies that the MIL lamp and circuit are operational. Refer to vehicle-specific information for a detailed description of MIL operation.

A DTC that is set in the ECM indicates that a certain test has been run and failed. A diagnostic flow chart for the specific DTC will generally aid the technician in diagnosing the cause of the DTC. Included with the diagnostic chart should be information about the DTC, such as the conditions necessary to run the test (also known as enabling criteria), what conditions cause the DTC to set, and the conditions necessary for the test to pass and clear the DTC.

8. Interpret OBD-II scan tool data stream, diagnostic trouble codes (DTCs), freeze frame data, system monitors, monitor readiness indicators, and trip and drive cycle information to determine system condition and verify repair effectiveness.

A scan tool communicates with the ECM and the vehicle's other on-board computers. The scan tool reads and displays diagnostic information provided by the ECM, such as DTCs and system operating data, or *parameters*.

Data transmitted to the scan tool from the ECM include both digital and analog values. Digital parameters are often called *switch parameters*, or *signals*, and are either on or off, high or low, yes or no. An analog parameter provides a signal value with a specific minimum-to-maximum range. These kinds of data include analog voltage readings, speed signals, temperature readings, and frequency ranges.

Every item of data transmitted from the ECM to a scan tool has specific value or signal range described in the vehicle specifications. These scan tool readings must be compared to specs to identify a system fault.

Scan tool readings that identify an open or a short circuit are among the easiest to recognize. If a resistive sensor displays a scan tool reading at or near the 5.0-volt reference voltage on which most such sensors operate, the sensor circuit to the ECM is open. If the scan tool voltage for such a sensor is at or near 0 volt, the circuit is probably grounded.

Scan tool readings of ECM input and output signals reflect values as processed by the ECM. In some systems, a sensor failure will cause the ECM to ignore the signal from the failed sensor and operate on back-up values stored in its own memory. In this case, the ECM may transmit the back-up values to the scan tool in place of the failed sensor signal. If any particular scan tool data reading does not make sense in relation to a particular problem or symptom, the system or component should be tested directly with a voltmeter, ohmmeter, oscilloscope, frequency counter, or other test equipment.

OBD-II diagnostic standards require a uniform library of diagnostic trouble codes (DTC) to be used by all carmakers. Additionally, all OBD-II systems must transmit a basic list of 16 data items to a generic scan tool. However, carmakers are free to "enhance" the on-board diagnostic capabilities of their control systems. Many systems provide much more than the minimum OBD-II diagnostic data requirements.

The OBD-II system monitors virtually all emission control systems and components that can affect tailpipe or evaporative emissions. In most cases, malfunctions must be detected before emissions exceed 1.5 times the emission standards. If a system or component exceeds emission thresholds or fails to operate within a manufacturer's specifications, a DTC will be stored and the MIL will be illuminated within two driving cycles.

The OBD-II system monitors for malfunctions either continuously, regardless of driving mode, or non-continuously, once per drive cycle during specific drive modes. A DTC is stored in the ECM when a malfunction is initially detected. In most cases the MIL is illuminated after two consecutive drive cycles with the malfunction present. Once the MIL is illuminated, three consecutive drive cycles without a malfunction detected are required to extinguish the MIL. The DTC is erased after 40 engine warm-up cycles once the MIL is extinguished.

In addition to specifying and standardizing much of the diagnostics and MIL operation, OBD-II requires the use of standard communication links and messages, standardized DTCs and terminology. Examples of standard diagnostic information are *freeze frame data* and Inspection Maintenance (IM) Readiness Indicators.

Freeze frame data consists of parameters such as engine RPM and load, state of fuel control, spark, and warm-up status. This information is stored in the ECM at the point the malfunction is initially detected; however, previously-stored conditions will be replaced if a fuel or misfire fault is detected. This data is accessible with the scan tool to assist in repairing the vehicle.

OBD-II Inspection Maintenance (IM) Readiness indicators show whether all of the OBD-II monitors have been completed. Test results can be displayed to show whether the monitor passed or failed. Checking to see if the monitor passed after a repair is a great way to verify that the repair fixed the vehicle.

> *Note:* Scan tool data includes data stream, diagnostic trouble codes, freeze frame data, system monitors, and readiness monitors.

▨ 9. Determine the relative importance of displayed scan tool data.

ECM serial data transmitted to a scan tool may have a data list of 50 or more items. Not all items of serial data are relevant to a single problem or symptom. Later sections of this overview discuss the most important serial data parameters for ignition, fuel, and emission system troubleshooting. For general powertrain control diagnosis, the following data parameters are among the most important:

- *System voltage*—The battery and charging system must provide a continuous regulated voltage of 12 to 14 volts. More importantly, the ECM must receive this voltage. Most serial data streams provide a reading of system voltage at the ECM. It must be within the normal range for the ECM and all system components to operate properly.

- *Engine speed, or RPM*—The engine rpm signal is the single most important input to the ECM. Without this signal, many engines will not even start. The RPM signal tells the ECM whether the engine is cranking, idling, accelerating, decelerating, or cruising. The RPM signal affects all ignition, fuel, emission, and transmission control operations of the ECM.
- *Vehicle speed, or mph*—Vehicle speed affects transmission control, ignition timing, fuel metering, and several emission-control subsystems. It is also an input signal for some antilock brake systems (ABS) and cruise control.
- *Temperature parameters*—The temperature of engine coolant, intake air, and transmission fluid affects fuel, ignition, and emission system operation, as well as transmission control.
- *Manifold pressure (or vacuum) and barometric pressure*—These air pressure parameters are primary inputs for fuel and ignition control and transmission shifting. Most control systems use air pressure measurements as a starting point for the ECM to calculate relative engine load.

For any given problem, only a few data parameters may be of primary importance for diagnosis. The ones listed above are a few that affect several subsystems or overall system operation. Others are described in later sections of this overview. Many scan tool manufacturers group related parameters together on their instrument displays or provide submenus of selected key data readings for specific subsystems. The most important parameters for overall system operation are often displayed at the top of a scan tool data display.

Control systems that comply with OBD-II on-board diagnostic standards provide a minimum list of operating data parameters. These include most of the important items to evaluate for general system troubleshooting. Basic OBD-II data parameters include: O_2 sensor voltage, open and closed-loop indications, engine speed and temperature, barometric and manifold pressure, air flow, short-term and long-term fuel trim corrections, intake air temperature, spark advance, engine load, and vehicle speed.

10. Differentiate between electronic powertrain control problems and mechanical problems.

Any given symptom or driver complaint can be caused by a problem in more than one system. A problem can appear to be based in one system when, in fact, it originates in another. For example, a longer-than- normal fuel injector pulse width and high long-term fuel trim correction factors are fuel system symptoms, but their causes can lie in several other engine subsystems. A mechanical problem, such as an intake vacuum leak, could be the root cause. On the other hand, an incorrect input signal from a barometric pressure sensor or manifold pressure sensor could cause these unusual fuel control symptoms.

Differentiating between mechanical problems and computer system problems is part of identifying and repairing the root cause of any symptom. A diagnostic technician must look for abnormal test results and isolate the cause of a problem to one system.

Information is often the key in diagnosing or repairing late-model computer-controlled systems. Diagnostic information will be available both in printed form and in computerized information retrieval systems in the shop or on the Internet.

Additionally, some equipment companies provide databases of test procedures and specifications that are built into their test equipment, which the technician can access while using the equipment to perform diagnostics. This is an emerging and very useful technology that gives the technician information when it is needed. Technical service bulletins (TSBs) are usually high-priority information for late-model systems and are included in all of the sources already mentioned. The technician must be aware of what service information is available and how to get the information needed. The best service information in the world is useless if the technician cannot readily retrieve the information.

Intermittent engine control problems will require diagnostic procedures tailored to the conditions present when the problem occurs. This service information needs to be gathered by carefully questioning the vehicle's owner so that an effective diagnosis may be performed. Heating or cooling affected components could indicate temperature-related faults. Performing wiggle tests and monitoring circuit values can find harness and connection problems. Problems such as these can go undetected if the technician is not provided with accurate service information.

11. Diagnose no-starting, hard starting, stalling, engine misfire, poor driveability, incorrect idle speed, poor idle, hesitation, backfire, surging, spark knock, power loss, reduced fuel economy, illuminated MIL, and emission problems caused by failures of computerized powertrain controls.

The proper operation of electronic powertrain controls is essential for the modern power plant to function correctly. A computer that loses power or ground will not function at all and will cause a no-start condition on any fuel-injected engine. All of the conditions listed in Task 11 have the potential to be traced back to the computer and its input and output components.

While a failure in the computer-control system is a possible source of these conditions, it should not be considered the problem until basic mechanical and electrical tests have been performed. A manifold absolute pressure sensor (MAP) will not generate a correct signal if the engine is incapable of producing normal manifold vacuum. An out of calibration coolant sensor input can cause hard starting when the engine is both cold and hot and can also alter fuel delivery enough to cause an emission test failure and poor mileage. A MAP or mass airflow (MAF) sensor that is out of calibration can cause many different driveability problems on a computer-controlled engine because it is a primary sensor input. Fuel delivery, spark timing, or torque converter clutch lock-up can all be affected by the MAP or MAF input.

Driveability concerns, such as hard starting, stalling, surging, spark knock, poor mileage, and excessive emissions, can all be traced to an incorrect MAP or MAF sensor signal. A problem with the throttle position sensor input may not only cause engine hesitation but may also affect transmission shift points or feel on electronically controlled transmissions. It is important to review the theory of operation section in the repair information for the vehicle to fully understand how each engine sensor input can affect the operation of the engine or transmission.

A failed output actuator, such as an EGR solenoid, may prevent the computer from properly controlling EGR gas flow. The computer may energize the EGR solenoid to allow vacuum to reach the EGR valve, while other systems energize the solenoid to block vacuum to the valve. In the first case, if the EGR valve does not receive vacuum, the engine may produce spark knock. In the second case, if the computer cannot limit EGR flow, hesitation, surging, and power loss may result. In both cases, understanding system operation is the key to correct diagnosis. A failure inside the computer, such as an open driver transistor for a fuel injector or ignition coil, can cause engine misfire and power loss. Computers may also exhibit intermittent driveability concerns caused by problems such as solder joint connections that are temperature or vibration sensitive.

These types of failures can be very challenging to diagnose and require employing advanced diagnostic procedures, such as circuit monitoring with such tools as labscopes and graphing multi-meters. The table below lists a few common problems that can be associated with a specific symptom. The table is not intended to be all-inclusive but is meant to point out how a component failure can affect the vehicle.

Symptom	Possible Cause
No Start	Open battery cable connection; fuel pump relay or circuit; ignition primary circuit fault
Hard Starting	Incorrect engine sensor input; low fuel pressure; weak spark; incorrect timing
Engine Misfire	DIS coil or wire failure; fouled or cracked spark plug; open or shorted fuel injector
Reduced Fuel Economy	Jumped timing belt or chain; restricted fuel filter; weak fuel pump; failed oxygen sensor; high fuel pressure; leaking fuel-pressure regulator
Incorrect Idle Speed	Dirty throttle body; defective idle speed control motor
Poor Idle	Vacuum leak; air intake leak behind MAF sensor
Hesitation	Bad throttle position sensor; dirty Mass Airflow Sensor element; binding Vane Airflow meter
Surging	Improper EGR operation; low fuel volume
Spark Knock	Over-advanced ignition timing; inoperative EGR valve; lean air/fuel mixture

12. Diagnose failures in the data communications bus network; determine needed repairs.

There are a number of different modules on today's vehicles that perform many jobs. These modules typically communicate with one another over a network or data bus. This network prevents the need for hardwiring one sensor to several different modules that use data from that sensor. The data bus can be either a one-wire or two-wire network that connects each module. There are many different designs and types of networks used on vehicles today, and many vehicles use more than one network. Before tackling any network fault, it is important to refer to service information to determine the type of network that is being diagnosed.

If one module fails to send needed information over the network, other modules may set a communications failure DTC for that module. The technician would then use the appropriate diagnostic chart for that DTC to diagnose why that module is not communicating properly over the network. A complete inspection of the power and ground circuits to the non-communicating module are generally performed when the cause of the problem is being diagnosed.

It is possible for one malfunctioning module to bring down the entire network and prevent any communication between modules. This is generally one of the more difficult problems to diagnose. One symptom of this fault would be no scan tool communication with any module on the network. When referring to the diagnostic chart for this condition, it is common to unplug each module on the network while checking for scan tool communication. When the module at fault is taken off of the network, the other modules will be able to communicate normally.

Physical damage to the network wiring can also cause the network to fail. These faults occur when the network wiring is open or shorted. The symptoms can vary depending on the location of the short or open and the type of network used. It is important to refer to the vehicle service information for guidance when diagnosing vehicle networks.

13. Diagnose failures in the anti-theft/immobilizer system; determine needed repairs.

The anti-theft system was designed to prevent the vehicle from being started in the event of theft. Many anti-theft systems involve special keys with chips or a transponder in the key. This prevents the vehicle from starting when any device other than the expected key attempts to turn the ignition switch. Other systems may have sensors in the ignition lock cylinder or switches in the door latches to detect tampering. Many newer vehicles are utilizing remotes that transmit unique signals and do not require a physical key at all. Regardless of the system, a malfunction usually produces the same symptom: The vehicle either starts and dies or simply does not start or crank at all. Refer to the vehicle's service information for a complete description of how the anti-theft system operates and a list of components included in the system. While there, be sure to check the normal operation of the security indicator and how the indicator will react to a fault or tampering in the anti-theft system. Compare normal operation of the indicator with the indicator on the vehicle. This will help determine if the condition is a result of the anti-theft system or a totally separate issue. A scan of the anti-theft module or body control module may produce DTCs that will aid in the diagnosis of the system. Some anti-theft diagnostic charts are symptom-based and will not provide any DTCs to aid in diagnosis.

14. Perform voltage drop tests on power circuits and ground circuits.

Voltage drop tests have always been important diagnostic methods for any electrical system. Voltage drop tests are typically used to check the integrity of a wire or connection and are very important in troubleshooting electronic control systems.

To perform a voltage drop test, the circuit must be energized (turn on the ignition or close the appropriate switch) to ensure that current is flowing in the circuit. Without current flow, voltage cannot drop across all the loads in the circuit. Place the leads of the voltmeter at each end of the wire or connection being tested. Lower current circuits should have very low voltage drops. Higher current circuits typically allow for larger voltage drops (up to 0.5 V on a starter cable).

Use the following values for maximum voltage drops across circuit wiring, connections, and switches:

- 0.00 V across a connection
- 0.20 V across a length of wire or cable
- 0.30 V across a switch
- 0.10 V at ground or across a ground connection

Because electronic systems operate with low voltage and very low current, a clean ground connection with minimum voltage drop is essential. To measure voltage drop across the complete ground side of a circuit, connect the voltmeter negative lead directly to the negative battery terminal. Then probe with or connect the positive meter lead to the ground circuit of the system being tested. The voltage drop of the ground circuit should not exceed 0.10 volt.

Remember that the sum of all the voltage drops in a circuit must exactly equal source voltage. If the measured voltage drops of all the designed circuit loads are less than source voltage, some unwanted resistance exists in the circuit. This is usually a damaged connector or wiring, corrosion, or a poor ground connection.

15. Perform current flow tests on system circuits.

Performing current flow tests through system circuits will require a Digital Multi-meter (DMM) and/or lab scope. Here is an example of how a current flow test should be performed on a canister purge circuit and what readings should be obtained:

1. Remove the fuse for the canister purge solenoid.
2. Connect an ammeter in place of the fuse.
3. Using a scan tool, command the solenoid on.
4. The reading on the ammeter should increase.

> *Note:* A typical good value should show an amperage increase of 400–700 mA when the solenoid is commanded on. With the solenoid off, amperage should decrease 400–700 mA. It should also be noted that the fuse may power several other components, and the meter may display a current reading with the solenoid off.

Current testing is required whenever a computer that controls an output device is being replaced. Output devices include items such as EGR valves, canister purge or shift solenoids, fuel injectors or the control circuit coil in a relay. A low resistance (shorted) solenoid or coil will increase current flow and possibly destroy a driver transistor in the computer. If the computer is replaced without identifying the defective solenoid or coil, the replacement computer will suffer the same failure. Current testing output devices should be done with a DMM set to the amps scale and connected with the meter in series so that the current flows through the meter. The reading should be taken over a five-minute time span to allow the device to heat up, and the value should not exceed manufacturer's specifications, usually less than 750 milliamps. Low resistance solenoids, such as pulse width modulated (PWM) transmission solenoids or fuel injectors that use current limiting drivers (Peak and Hold), cannot be tested in the above manner due to excessive current flow. A labscope and current probe or resistance testing with an ohmmeter will be required.

16. Perform continuity/resistance tests on system circuits and components.

When performing diagnostic checks, the technician is commonly required to measure resistance and continuity with a meter. When checking a wire for continuity, it should have little to no resistance. Most multi-meters have a continuity function. One lead of the meter is placed at each end of the wire to be tested. When testing for continuity, be sure that there is no voltage on the circuit and that it is isolated from solid-state modules that could be damaged by the meter. The circuit should also be isolated from other components or circuits that could interfere with test readings. If the meter beeps, this indicates a good path for conductive flow, and the wire may be considered good. Circuits may also be tested for continuity to ground in the event of a short to ground. The circuit should be isolated, as indicated before, with one lead on the wire and the other connected to chassis ground. A beeping meter indicates that the circuit is shorted to ground. On the other hand, a continuity test can be used on ground circuits to check the integrity of the ground connection. In this case, a beeping meter indicates that the ground connection is good. A continuity test should not replace a voltage drop test, which should be used in circuits that carry any amount of current.

Resistance checks of circuits and components are also commonly made during diagnostic procedures. Always refer to the resistance specification of the component before testing. Resistance checks are made with the circuit inoperative and isolated from modules or other circuits that could interfere with resistance readings. When comparing resistance readings with specs of a component, a reading lower than spec indicates that the component is shorted, while a reading higher than spec indicates a high internal resistance. A meter reading of "O.L" could indicate that the component is open or that the meter is not in the proper range. The use of an auto ranging meter will help avoid this issue.

17. Test input sensor/sensor circuit using scan tool data and/or waveform analysis.

Signal analysis using a scan tool, labscope or graphing multimeter is becoming one of the most important skills a driveability technician can possess today. A good technician should be able to look at a signal or waveform and determine whether or not it is acceptable. This means that the technician should understand what both good and bad signals or waveforms look like. Many of today's automotive digital storage oscilloscopes (DSOs) allow a technician to save screens and store them in a computer software program for retrieval. These programs provide the means to build a database of good and bad waveforms and can also print out waveforms for the customer in order to show repair verification. Some equipment is also equipped with a stored memory of confirmed good waveforms that may be used for comparison with the signal under test. Voltage testing an input signal alone does not provide the level of detail necessary to diagnose problems in the complex engine control systems used today. The DSO can pinpoint signal dropouts and glitches far better than the best DMMs can.

18. Test output actuator/output circuit using scan tool, scan tool data and/or waveform analysis.

Testing procedures for output actuators using scan tools and waveform analysis can be enhanced if the system supports bi-directional testing. Using the scan tool's bi-directional controls, many computer outputs can be turned on and off by the technician, and measurements can be taken with a DSO. Even without bi-directional control, the DSO can confirm output actuator operation based on computer commands monitored through scan tool data. Checking signals right at the ECM connector is generally the best method to determine if a circuit is operating properly. For instance, the DSO can show a canister purge or TCC solenoid turning on and off as commanded during a test drive while the scan data parameter for that circuit is simultaneously being checked with the scan tool.

19. Confirm the accuracy of observed scan tool data by directly measuring a system, circuit, or component for the actual value.

Using direct measurement to confirm accuracy of displayed scan tool data generally involves performing tests to determine calibration errors. An infrared temperature-measuring tool may be used to check cylinder head temperature where the coolant sensor is mounted while also viewing the sensor scan parameter on a scan tool. Too large a discrepancy between the two readings means the coolant sensor is out of calibration. A similar test is performed on a manifold pressure sensor using a hand-held vacuum pump to supply vacuum and a DMM to measure the sensor signal voltage. Two readings are taken at different vacuum levels, and one reading subtracted is from the other to determine proper sensor calibration. Performing this type of testing is crucial to eliminating incorrect diagnosis of driveability problems and unnecessary parts replacement.

20. Test and confirm operation of electrical/electronic circuits not displayed in scan tool data.

Voltage drop testing discussed under task 14 is the most common test principle applied to electrical and electronic components for which serial data is not available. Actuator test mode (ATM) tests or output state checks can also be used to analyze the operation of output devices for which serial data is not available.

Verifying the B+ voltage supply to the ECM and to other system components is a basic part of troubleshooting. The ECM and output devices for which the ECM provides ground control receive battery voltage from the vehicle electrical system. Many system sensors receive a 5.0-volt reference voltage from the ECM.

The best way to check any supply (B+) voltage is with a high-impedance digital voltmeter. A simple probe light should not be used because it only indicates that some voltage is present; it does not indicate the actual voltage level. Also, a probe light may draw excessive current through a circuit branch and damage electronic components.

Along with verifying the voltage supply to system devices, proper ground connections must also be verified. An open ground connection can be identified by simple inspection or by measuring open-circuit voltage at a point in a circuit where there should be a voltage drop. A high-resistance ground is best identified by voltage drop measurement.

An ammeter can be used to verify and measure current in an operating circuit, and an ammeter or a voltmeter, or both, can be used to check the operation of solenoids, relays, and motors.

Another method used to determine if a component is operating is by simply listening to the component. Most solenoids will make some kind of noise when energized. Stethoscopes are commonly used to listen to injectors clicking to verify their operation.

21. Determine root cause of failures.

22. Determine root cause of multiple component failures.

23. Determine root cause of repeated component failures.

Tasks 21, 22, and 23 are related to each other and to similar tasks listed for other subsystems. An example of these principles applied to the control system in general is the case of ECM replacement because of damaged output driver transistors. If an ECM is replaced because of transistor failure, the cause of the transistor damage must be determined, or the replacement unit may become damaged as well.

ECM driver transistors are often damaged by short circuits in output solenoids, relays, motors, and fuel injectors. A short circuit in one of these devices can increase current flow through its ECM transistor and lead to failure. Identifying and replacing the defective output device isolates the root cause of the failure and avoids repeated failures.

A case history exists of an ECM with transistor damage that was replaced without identifying the root cause of the problem, which was a shorted fuel injector. This short circuit in the injector allowed the engine to run overly rich, which not only took out the new ECM but also eventually melted down the catalytic converter. Several high dollar components were destroyed because one small injector failed.

24. Verify effectiveness of repairs.

After repairs have been made, steps should be taken to verify the effectiveness of the repair. Re-perform any tests that were made during diagnosis to verify that the system passes. Refer to any scan tool data that will ensure proper system operation. If the condition set a DTC, refer to the enabling criteria for that particular DTC and operate the vehicle under those specific conditions so that the test will be run. Refer to any diagnostic monitors that failed prior to the test and confirm that they have run and passed. If possible, drive the vehicle through at least two drive cycles and confirm that the MIL lamp is not illuminated. During test driving, the vehicle should be operated in the same manner as when the fault occurred before repairs were made. Check OBD-II vehicles for pending codes that may have failed once but have not yet caused the MIL to illuminate.

C. Ignition System Diagnosis (6 Questions)

1. Verify customer concern; determine if the concern is the result of a malfunction or normal system operation.

Before diagnosis can begin, the customer's concern must be verified. Knowledge of vehicle operation and on-board diagnostics will help the technician determine if the concern is normal system operation or the result of a malfunction. Refer to service information and research the normal operation of the vehicle. Accurate descriptions are available to familiarize the technician with normal system operation. Apply this information to the customer's concern to determine if a malfunction exists. If the concern has been duplicated and determined to be a malfunction, note the conditions present that caused the concern to occur, such as the amount of time the vehicle has been driven, engine temperature, engine load, RPM, vehicle speed, etc. Identify any abnormal scan tool data readings that may help in diagnosis of the vehicle. Keep track of these conditions and readings, as they will aid in verifying the effectiveness of the repair.

2. Inspect and test for missing, modified, inoperative, or tampered components.

Inspect the vehicle's ignition system for missing components such as heat shields or protective conduit. Should these items become missing, ignition components could become burned or damaged, which could result in a misfire. Always be sure that these components are installed to prevent repeat failures.

Check to see that the vehicle has the appropriate spark plugs installed. Using plugs that have non-recommended styles of electrodes or of a different heat range can affect emissions levels and vehicle performance.

Since much of the ignition system today is electronic and cannot be adjusted, tampering with the ignition system is not as easily done on today's vehicles. One area that can be tampered with is the software programmed into the ECM. Ensure that the software present is supported by the manufacturer.

3. Locate relevant service information.

Service information for the ignition system can come from the same sources as other information about the car. Among the information or service data you will need for the ignition system are the following:

1. Correct spark plug part numbers, including heat range and resistor requirements
2. Spark plug gap and installation torque
3. Ignition cable (spark plug wire) resistance

Computer-controlled spark timing systems use various sensors to indicate crankshaft position and to identify cylinders by number. A technician must know the number and kinds of ignition control sensors on a particular system for accurate troubleshooting.

As with other control system components and subsystems, technical service bulletins (TSB) are important sources of ignition system service specifications and diagnostic instructions.

4. Research system operation using technical information to determine diagnostic procedure.

A technician should develop a diagnostic routine based on a thorough understanding of system operation. Before starting a diagnosis, the technician should research system operation in the service material that is available to them. Most computer-based information systems, as well as manufacturers' printed manuals, begin the service section with an overview of the system operation so that the technician is familiar with the normal operation. This material should not be overlooked. Service information will usually include symptom-based diagnostic procedures to help guide the technician in performing the needed tests to diagnose a complaint. Knowing what type of primary triggering device is used and whether the signal goes to an ignition control module (ICM) or directly to the ECM is necessary information when determining a diagnostic approach to a no-start condition.

5. Use appropriate diagnostic procedures based on available vehicle data and service information; determine if available information is adequate to proceed with effective diagnosis.

Some of the simplest ignition system checks are resistance tests of spark plug cables and ignition coils with an ohmmeter. If the ohmmeter indicates infinite resistance, it can be assumed that an open circuit exists in the component. For complete testing, however, the manufacturer's resistance specifications must be known. Without these specifications, a low-resistance or high-resistance condition that may be leading to a short-circuit or an open-circuit problem cannot be pinpointed.

Service information that describes the type of engine sensors and their voltage waveform signals will help the technician choose diagnostic procedures for those specific sensors. Without test procedures and waveform information, it may be possible to determine if voltage is present or absent in a circuit, but it will not be possible to tell if the signal is being received and understood by the ignition module or the ECM.

6. Determine the relative importance of displayed scan tool data.

The following data parameters are among the most important for ignition system troubleshooting and for determining ignition effects on other subsystems:

- *Engine speed, or RPM*—The engine RPM, or *tach* signal tells the ECM whether the engine is cranking, idling, accelerating, decelerating, or cruising. The tach signal affects all ignition control operations of the ECM.
- *Crankshaft position (CKP) signal*—The CKP signal usually provides the engine RPM information to the ECM, and it indicates the angular position of the crankshaft and provides cylinder identification for spark advance control.
- *Spark advance*—Many systems provide direct datastream readings of spark advance.
- *Manifold pressure (or vacuum) and barometric pressure*—These air pressure parameters are primary inputs for ignition control. The MAP sensor in a late-model engine takes the place of the vacuum advance diaphragm on an older distributor.
- *Vehicle speed*—Vehicle speed affects ignition timing and several other control subsystems.
- *Temperature parameters*—The temperature of engine coolant and intake air affects ignition operation.
- *Throttle position*—The TPS signal has an immediate effect on electronic spark timing as the ECM changes spark advance for acceleration and deceleration, as well as at idle.

Control systems that comply with OBD-II on-board diagnostic standards must provide data parameters for engine speed, spark advance, manifold pressure, throttle position, and coolant temperature as part of the minimum list of operating data parameters.

7. Differentiate between ignition electrical/electronic and ignition mechanical problems.

Electrical problems can mimic mechanical problems and vice versa. In an ignition system, intermittent problems in circuits between the ignition control module (ICM) and the ECM can cause erratic spark timing and engine misfire. The same performance symptoms can also result from mechanical damage to ignition sensors like the CKP sensor. To distinguish between an electrical or mechanical fault as the root cause of a problem, the technician must inspect system components for obvious damage and perform electrical tests to identify circuit problems.

Certain mechanical problems can have a negative effect on the ignition system. Improperly adjusted valves, or any other mechanical defect that makes noise, can be picked up by the knock sensor. The ECM sees this as spark knock and retards engine timing. Engine performance decreases as emissions increase.

Faulty or damaged crank sensor trigger wheels can also cause ignition faults. When the wheel is damaged, the ECM may interpret this as a change in engine speed and set misfire DTCs for two of the engine's companion cylinders.

8. Diagnose no-starting, hard-starting, stalling, engine misfire, poor driveability, backfire, spark knock, power loss, reduced fuel economy, illuminated MIL, and emission problems caused by failures in the electronic ignition (EI) systems; determine needed repairs.

Some tests for EI systems are unique to certain kinds of systems; others are basic checks that apply to all ignition systems. Battery voltage must be supplied to one side of the primary windings for the coils, and switching control must be provided on the ground side. Absence of battery voltage or primary switching can cause hard starting or a no-start condition. Similarly, open circuits in spark plug cables and ignition coils can cause the same problems.

When misfiring, detonation, poor performance, and increased emissions are related to the ignition system, they often result from spark timing problems. In an EI system, the ECM controls timing. Timing problems can result from erratic sensor input signals and circuit problems between the ECM and the ICM. Some traditional causes of misfiring include breakdown of the ignition coils under load, open or shorted spark plug cables, and worn or damaged spark plugs.

The knock sensor input is an important feature of most computer-controlled ignition systems. When this sensor detects engine detonation, it signals the ECM to retard ignition timing (or decrease the advance) and increase EGR flow until the detonation stops.

9. Test for ignition system failures under various engine load conditions.

The ignition system's *available voltage* is the maximum secondary voltage that the system can deliver. The *required voltage* is the secondary voltage necessary under any condition to fire the spark plug. The required voltage varies with changes in engine speed and load, but available voltage must always be greater than required voltage. The difference between the two is the *voltage reserve*.

Testing for ignition faults under engine load is the process of determining available voltage, required voltage, and voltage reserve. Problems that can lower the available voltage include:

- Less than the required primary voltage supplied to the coil primary windings.
- High resistance anywhere in the coil primary circuit.
- Shorted primary or secondary windings in the coil.
- Open or high resistance in the primary or secondary windings of the coil.

Some problems that can raise the required voltage include:

- Open circuits or excessive resistance in the spark plug wires.
- Worn spark plugs or wide plug gaps.
- Lean air/fuel mixtures.
- Excessive carbon deposits in combustion chambers that increase cylinder pressure and temperature.

Any of these problems can cause ignition misfire under load if the required voltage is greater than the available voltage. Problems that create a lower resistance path to ground in the secondary circuit will also cause misfire. Spark plug wire insulation leakage, torn spark plug boots, or cracked spark plug insulators will all allow secondary voltage to leak to ground rather than jumping the spark plug gap. Tests such as power braking (brake torque or stall testing) the engine or cranking KV tests will stress the secondary circuit and identify these types of faults.

10. Test ignition system component operation using waveform analysis.

The use of the lab scope gives the technician a graphed visual picture of the ignition system component's activity. Technicians should have an understanding of the equipment being used and the different types of patterns they will encounter while testing this system. The technician will be testing primary, secondary, coils, wires, crank sensors, cam sensors, and spark plugs. Faulty spark plug wires commonly cause high secondary resistance and will appear as a high firing line on the scope.

11. Confirm ignition timing and/or spark timing control.

The simplest distributor-type electronic ignitions have centrifugal and vacuum advance mechanisms. You can check base timing and spark advance as it you would on an older breaker-point system. Electronic spark timing (or computer-controlled spark advance) eliminated centrifugal and vacuum advance, but distributor systems still require base timing test and adjustment. To check base timing on these later systems, you must take the computer out of the timing control loop by disconnecting or jumping a specified connector. Spark advance with computer-controlled spark timing generally is not tested. Distributorless eElectronic ignitions do not require a base timing test or adjustment. All timing adjustments are performed by the Engine Control Module (ECM). However, spark advance, retard values and knock sensor activity can still be viewed on a scan tool. Comparing these values with known good values will let the technician know the timing needs of the engine.

12. Determine root cause of failures.

13. Determine root cause of multiple component failures.

14. Determine root cause of repeated component failures.

Tasks 12, 13, and 14 are related to each other and to similar tasks listed for other subsystems. An example of these principles applied to the ignition system is the case of an ignition module that fails repeatedly because of high temperature. High temperature can be caused by excessive current draw due to shorted ignition coil primary windings. High temperature faults can also result from high underhood temperatures. Engine overheating, missing heat shields, or high ambient temperatures can be the root cause of repeated module failures.

Spark plugs that foul after very few miles of operation are a classic example of a root cause located in another engine system. In this case, mechanical failures like broken rings or worn valve guides can cause the plug to foul, and new plugs will continue to foul until the root cause is repaired.

15. Verify effectiveness of repairs.

After repairs have been made, steps should be taken to verify the effectiveness of the repair. Re-perform any tests that were made during diagnosis to verify that the system passes. Refer to any scan tool data, such as misfire counters, that will ensure proper system operation. If the condition set a DTC, refer to the enabling criteria for that particular DTC and operate the vehicle under those specific conditions so that the test will be run. Refer to any diagnostic monitors related to the ignition system that failed prior to the test and confirm that they have run and passed. If possible, drive the vehicle through at least two drive cycles and confirm that the MIL lamp is not illuminated. During test driving, the vehicle should be operated in the same manner as when the fault occurred before repairs were made. Check OBD-II vehicles for pending codes that may have failed once but have not yet caused the MIL to illuminate.

D. Fuel Systems and Air Induction Systems Diagnosis (8 Questions)

1. Verify customer concern; determine if the concern is the result of a malfunction or normal system operation.

Before diagnosis can begin, the customer's concern must be verified. Knowledge of vehicle operation and of on-board diagnostics will help the technician determine if the concern is normal system operation or the result of a malfunction. Refer to service information and research the normal operation of the vehicle. Accurate descriptions are available to familiarize the technician with normal system operation. Apply this information to the customer's concern to determine if a malfunction exists. If the concern has been duplicated and determined to be a malfunction, note the conditions present that caused the concern to occur, such as the amount of time the vehicle has been driven, engine temperature, engine load, RPM, vehicle speed, etc. Identify any abnormal scan tool data readings that may help in diagnosis of the vehicle. Keep track of these conditions and readings, as they will aid in verifying the effectiveness of the repair.

2. Inspect and test for missing, modified, inoperative, or tampered components.

Inspect throttle bodies for missing caps or plugs on the throttle adjustment screws. Also look for any other attempts to adjust idle speed other than by original-equipment means.

Air induction systems must meet original-equipment specifications and have all components that were originally installed on the car. Modified air intake systems can increase HC and CO emissions, particularly at cold and hot temperature extremes. Oil type air filters can contaminate mass airflow (MAF) sensors and lead to a lean condition.

Fuel-injection systems are less susceptible to tampering, but unapproved devices and adjustments are still used. Some items to check for include aftermarket performance parts, such as computer program changes, adjustable fuel-pressure regulators, high flow fuel injectors, intake manifolds or modified airflow sensors. The entire air intake system must be in place and functional. Fuel pumps, filters, and delivery systems must also meet original-equipment specifications. Fuel delivery systems are interconnected with fuel evaporation emission systems. In some cases, modifications to the fuel delivery system can increase evaporative emissions.

3. Locate relevant service information.

Service information for the fuel and air intake systems comes from the same sources as other information about the car. Whether it be printed or from a computer-based system, service information should include normal system operation, system specifications, and testing procedures. Technicians should also familiarize themselves with the types and number of sensors used to control fuel and air delivery to the vehicle.

A technician should determine things like whether the ECM uses a mass airflow (MAF) sensor to determine the engine's fuel needs, or if the ECM uses a manifold absolute pressure (MAP) sensor for that task. Refer to the number and locations of oxygen (O_2) or air fuel ratio sensors that are used.

If the vehicle is equipped with electronic throttle control, it may be wise to find information for that system, as it can vary in its operation from one manufacturer to another.

4. Research system operation using technical information to determine diagnostic procedure.

For the technician diagnosing problems with fuel systems or air induction systems, the proper starting point must be with a thorough understanding of system operation. Using printed or electronic service information, the technician should research how the system is configured and what testing procedures the manufacturer suggests. A search of technical service bulletins should also be performed before detailed testing is started. Determining specifications (such as fuel pressure and volume), component locations (such as fuel pump relays or inertia switches), and air induction system components (such as intake manifold tuning valves or runner controls) will speed testing and diagnosis.

5. Evaluate the relationships between fuel trim values, oxygen sensor readings, air/fuel ratio sensor readings, and other sensor data to determine fuel system control performance.

Fuel trim is a diagnostic strategy that enables the ECM to adapt to changing engine conditions and ensure an air fuel ratio of 14.7:1 (*stoich*). Fuel trim values can also be used as a diagnostic tool used to determine the overall rich or lean status of the vehicle. Fuel trim is displayed as the percent of fuel that the ECM is either adding or subtracting to achieve a 14.7:1 air fuel ratio. Positive fuel trim values indicate that the ECM is adding more fuel than normal, which specifies a lean air fuel ratio. Vacuum leaks, low fuel pressure, clogged injectors, plugged fuel filters, or contaminated MAF or oxygen sensors can cause fuel trim numbers to increase. Negative values indicate that the ECM is taking fuel away from the engine, which signifies a rich air fuel ratio. High fuel pressure, leaking injectors or fuel pressure regulators, cooling system malfunctions, and inaccurate temperature sensor readings will all cause fuel trim numbers to decrease. If fuel trim values exceed the ECM's ability to maintain a stoichiometric air fuel ratio, a DTC is set and the MIL is illuminated.

Two types of fuel trim that are calculated are *Short Term* and *Long Term*. Short Term Fuel Trim (STFT) is typically calculated directly from exhaust feedback sensor readings, either oxygen (O_2) or air fuel (AF) ratio sensors. This is an instant fuel trim calculation based solely on the sensor readings. Exhaust feedback sensors that continually produce a lean reading will cause STFT numbers to increase. The ECM will increase the amount of fuel being delivered in an attempt to cause the feedback sensor to show the desired air fuel ratio reading. O_2 sensors should produce a voltage pattern that continually switches from below 250 mV to above 800 mV. AF sensors will normally show a voltage around 3.3 V and not switch. AF sensors may also display current. Refer to the vehicle's service information for the proper AF sensor reading and how to interpret the data shown. Long Term Fuel Trim (LTFT) is a calculation based not only on exhaust feedback sensors but on many of the engine's other sensors as well. LTFT displays a better overall average of the fuel trim over a period of time, along with the current engine operating conditions. Many technicians primarily use LTFT when diagnosing fuel-related issues.

6. Use appropriate diagnostic procedures based on available vehicle data and service information; determine if available information is adequate to proceed with effective diagnosis.

One of the most basic fuel system diagnostic sequences is to determine if the engine control system is operating in open loop or closed loop. The voltage signal from the O_2 signal is the primary indication of loop status, both to the vehicle ECM and to the technician.

After establishing whether the system is in open or closed loop, scan tool data and four- or five-gas exhaust analyzer readings can be evaluated to determine if the ECM is providing a rich or a lean correction and the extent of any such correction. The scan tool data items summarized under Task 7 are the primary indicators of both loop status and rich or lean operation. Determining these basic conditions of control system operation is the first step in troubleshooting the fuel subsystem as well as other areas of the engine control system.

7. Determine the relative importance of displayed scan tool data.

The following data parameters are the most important for fuel and air induction system troubleshooting and for determining the effects of these systems on other subsystems:

- *Engine speed, or RPM*—The engine RPM, or *tach* signal, tells the ECM whether the engine is cranking, idling, accelerating, decelerating, or cruising. The tach signal affects all fuel control operations of the ECM.
- O_2 Sensor—The exhaust oxygen sensor (O_2S) is the feedback signal that the ECM uses to control air/fuel ratios and is also the primary input that the ECM uses to determine open- or closed-loop fuel control operation. When the O_2 sensor measures excess oxygen in the exhaust, the fuel mixture is too lean, and the ECM will respond by making the mixture slightly richer. When the O_2 sensor measures less than normal oxygen in the exhaust, the fuel mixture is too rich, and the ECM will respond by making the mixture slightly leaner.
- *Loop status (open or closed)*—This is an internal ECM parameter that indicates the status of fuel mixture control. When in open loop, the ECM must use engine sensors such as MAP or MAF, TPS, RPM, ECT, and IAT to determine how much fuel the engine needs. Open loop operation generally occurs when the engine is first started, and the O_2 sensors are not warm enough to provide an accurate indication of air/fuel mixture. After the O_2 sensor has reached operating temperature, the ECM will enter closed loop. The ECM can use the O_2 signal to determine if the fuel mixture is rich or lean. When in closed loop, the ECM can get feedback from the O_2 sensor on any fuel mixture adjustments that are made.
- *Intake air mass*—All fuel-control systems that transmit serial data provide some indication of intake air measurement. Air measurement may be based on (1) speed density (manifold pressure and RPM), (2) air velocity (vane airflow sensor), or (3) mass airflow (air molecular mass, or weight). Data readings may be in voltage, frequency, or grams per second of airflow.
- *Long-term and short-term fuel trim correction*—Many fuel-injection systems provide data parameters that indicate the long-term trends and short-term actions to correct a fuel mixture in either the lean or the rich direction. OBD-II on-board diagnostic standards require that these values be given as percentages.

- *Idle speed control operation*—The engine RPM signal indicates actual idle speed. Most systems also provide data on desired idle speed and on the operation of idle air control (IAC) valves or throttle solenoids and motors.
- *Temperature parameters*—Engine coolant temperature (ECT) and intake air temperature (IAT) are two primary input factors that the ECM uses to determine the air/fuel ratio, especially when in open loop.
- *Throttle position*—The TPS signal has an immediate effect on fuel control as the ECM changes air/fuel ratio for acceleration, deceleration, cruising, and idle.
- *Vehicle speed*—Vehicle speed also affects fuel control.

Control systems that comply with OBD-II on-board diagnostic standards must provide data parameters for engine speed, loop status, fuel trim correction, intake air mass, throttle position, coolant and air temperature, and O_2 sensor readings as part of the minimum list of operating data parameters.

8. Differentiate between fuel system mechanical and fuel system electrical/electronic problems.

A vacuum leak is the most common example of a mechanical problem that affects fuel control and that will cause abnormal serial data readings and electronic control operation. Just as too much air entering the intake system through a leak will cause problems, too little leaving through the exhaust will cause its own set of problems. A restricted exhaust is a mechanical problem that will affect data for manifold pressure, fuel-injection pulse width, and O_2 sensor signals.

When diagnosing the intake system, an air leak in the intake duct or plenum may or may not affect fuel control. If the fuel-injection system is the speed density type that uses manifold pressure to calculate intake air density, an air leak in the intake duct may have minimal effects. The extra air entering the manifold is sensed and measured by the MAP sensor and accounted for in the air/fuel ratio. However, if the leak is great enough, the idle speed system may not be able to compensate for the excess air, and idle speed may be high. A vacuum leak could also change the symptoms, depending upon its location. For example, a vacuum leak upstream that affects all cylinders would affect the entire engine, while a vacuum leak at the intake runner next to the cylinder head would most likely affect only that cylinder.

If, on the other hand, the fuel-injection system uses a mass airflow (MAF) sensor, and the intake air leak is downstream from the MAF sensor, the extra airflow will not be measured. This kind of intake air leak can cause abnormal lean mixtures, rough or unstable idle, and may set a DTC for the fuel system.

9. Differentiate between air induction system mechanical and air induction system electrical/electronic problems, including electronic throttle actuator control (TAC) systems.

When dealing with air induction faults, the technician must first determine if the condition is a mechanical fault or is due to an electrical system or component. Mechanical air induction faults can include vacuum leaks, dirty air filters, loose or deteriorated vacuum hoses, dirty throttle passages, and cracked or leaking intake piping. Any of these faults will cause a variety of engine performance issues and can generally be identified with a thorough visual inspection. Electrical faults usually cannot be identified as easily. Electrical components or systems that can cause air induction problems can include faults in the variable induction system (such as an intake tuning valve or manifold runner control), the IAC or air bypass, the TAC system, and the MAF sensor. Scan tool data and actuator tests should be considered useful tools when diagnosing these components. Check for any DTCs that may be associated with the customer's concern, and follow diagnostic charts for these DTCs if present.

When dealing with a TAC system, be sure to locate a description of the systems operation. TAC systems usually have default modes of operation when a fault occurs. Most of these systems use dual TPS sensors and dual accelerator pedal position sensors. In the event of one sensor failure, the system may enter a failsafe mode that will only allow the throttle to be opened a limited amount. In the event of two sensor failures, the system may be disabled and the throttle plate set to a default position that sets idle speed to around 1500 rpm. Some vehicles use a separate malfunction indicator lamp specifically for the TAC system. Be familiar with this indicator so that it can be recognized when illuminated.

10. Diagnose hot or cold no-starting, hard starting, stalling, engine misfire, spark knock, poor driveability, incorrect idle speed, poor idle, flooding, hesitation, backfire, surging, power loss, reduced fuel economy, illuminated MIL, and emission problems on vehicles equipped with multiport <u>fuel injection</u> and direct injection fuel systems; determine needed action.

A major fuel-delivery problem for fuel-injection systems is incorrect fuel pressure, which will be discussed for Task 11. Incorrect fuel pressure can cause or contribute to many of the symptoms listed here.

Another major fuel injection problem that can cause many of these symptoms is improper fuel flow through the injectors. Restricted injectors can cause problems in hot or cold starting as well as all phases of operation. Restricted injectors tend to cause lean mixtures and emission problems related to lean air/fuel ratios. Leaking injectors can contribute to many of the same problems as restricted injectors. Leaking injectors, however, tend to cause rich mixtures and the fuel mileage and emission problems related to rich air/fuel ratios.

Vehicle manufacturers' service manuals often contain diagnostic flowcharts and procedures for each of these symptoms. This is another example of the importance of having complete and accurate diagnostic information.

11. Inspect fuel for quality, contamination, water content and alcohol content; test fuel system pressure and fuel system volume.

Verifying fuel quality is a step that should not be overlooked when diagnosing performance complaints related to the fuel system. Water in the fuel, stale fuel, or excessive alcohol content will all have an adverse effect on engine performance. Commercially-available test kits can be obtained that allow measurements to be made concerning fuel quality, such as alcohol content and volatility.

Most port fuel-injection systems have a pressure test port on the fuel rail. Many throttle body systems also have a test port. If the system has such a test port, it is normally fitted with a Schrader valve. Simply connect a test gauge to the port and crank or start the engine to check pressure. If the fuel-injection system does not have a test port, install the gauge on the pressure line with the appropriate adapters.

As a broad, general rule, port fuel-injection systems operate at higher pressures than throttle-body systems. Pressures for PFI systems can range from 28 to 65 psi; TBI system pressures range from 9 to 30 psi. These are only general limits, however; the carmaker's exact specifications must be obtained for accurate testing. If fuel-pressure values are used in a question on the test, the specifications required to make a diagnosis may be given in the Composite Vehicle reference.

Fuel pressure regulated by the vacuum regulator with the engine idling will be 8 to 10 psi lower than with no vacuum applied to the regulator. The vacuum regulator maintains a constant pressure drop across the fuel injector in response to changing intake manifold pressure present at the injector's tip. When intake manifold pressure rises, fuel pressure rises to keep fuel delivery constant. A fuel pump should produce pressure that is higher than the maximum regulated pressure. Momentarily clamping off the fuel return line or deadheading the pump can measure this pressure.

Some fuel-injection systems have fuel pump check valves and regulators that close to maintain system pressure at rest when the pump stops. This rest pressure should be checked. Any rapid drop in rest pressure typically indicates a problem and should be diagnosed by clamping off the feed and return lines alternately to isolate the leak. If the pressure remains constant with the feed line clamped, a leaking fuel pump check valve is indicated. It should be noted that some low-pressure TBI systems do not hold pressure when the pump is not running. Steady pressure with the return line clamped means a leaking pressure regulator, and if leak-down occurs with both lines clamped, an injector or regulator is leaking.

12. Evaluate fuel injector and fuel pump performance (mechanical and electrical operation).

Faulty fuel injectors can cause uneven idle speeds and surging or hesitation at cruising speeds. Emission test failures for high CO can indicate a rich condition caused by leaking injectors. Similarly, high HC readings can be caused by a lean misfire due to plugged injectors. High firing voltage and short duration spark lines on an ignition oscilloscope secondary waveform can indicate a lean mixture caused by restricted injectors.

On the car, testing of fuel injectors for flow rate is done with a pressure drop test. Each injector is operated one at a time with a special fixed rate pulse tool, while the pressure drop in the fuel rail is monitored with a gauge. A lower-than-average pressure drop indicates a lean or restricted injector, while a greater-than- average pressure drop indicates a rich or leaking injector. More accurate car injector test equipment can measure fuel flow in milliliters and also check spray patterns. This type of equipment will also clean the injectors during testing, and injector sets can be flow matched for optimal performance.

Injectors can be tested electronically with a digital storage oscilloscope. Both voltage and current waveforms can indicate problems with the circuit or the injector itself. Low inductive spikes seen on a voltage waveform can indicate a shorted injector coil. Fuel injector feed voltage and ground levels can be measured on a voltage waveform. A low amp current probe does an even better job of indicating shorted or high resistance injector coils and can also display injector pintle opening events in the waveform. This test can pick out sticking injectors that may otherwise be difficult to find. A low amp probe tests the injector coil under full voltage, dynamic conditions and is much more accurate than testing injector resistance with an ohmmeter.

Many late-model engine control systems for sequential fuel injection have built-in self-tests in which the vehicle ECM turns off each injector in sequence. The ECM then records the RPM drop and indicates on the data stream which injectors may not have provided a sufficient speed drop. This test is similar to the cylinder balance test provided on many engine analyzers, in which the secondary ignition is disabled cylinder by cylinder. It should be noted that the fuel-injection balance test only indicates that a cylinder is not delivering full power. The injector or other components may be the cause.

Evaluating fuel pump performance mechanical operation is done with a pressure test listed in Task 11. Even a mechanically-sound fuel pump can fail this test if the electrical circuit is faulty. Voltage drop tests on the fuel pump will insure that the pump has a proper power feed and a good ground connection. A DSO with a low amp probe can provide a current waveform that shows the integrity of the brushes and commutator in the motor as well as the current draw of the pump motor. Testing of known good systems to determine normal values will allow a technician to quickly pinpoint problems with a current probe and scope.

13. Determine root cause of failures.

14. Determine root cause of multiple component failures.

15. Determine root cause of repeated component failures.

Tasks 13, 14, and 15 are related to each other and to similar tasks listed for other subsystems. A well-documented problem is with repeated fuel pump failures. After several replacements of good quality fuel pumps, it is determined that debris in the fuel tank is getting into the fuel pumps and damaging the pump. After removing and properly cleaning the tank, the problem disappears. A sticking fuel-pressure regulator can increase fuel pressure high enough to damage the fuel pump. Failure to test fuel pressure after replacing the pump will prevent the technician from finding the root cause of the problem—a bad pressure regulator. Problems such as these illustrate the need to test the system after repairs are made instead of only testing during the diagnostic process.

16. Verify effectiveness of repairs.

After repairs have been made, steps should be taken to verify the effectiveness of the repair. Re-perform any tests that were made during diagnosis to verify that the system passes. Refer to any scan tool data, such as fuel trim and exhaust feedback sensors that will ensure proper system operation. If the condition set a DTC, refer to the enabling criteria for that particular DTC and operate the vehicle under those specific conditions so that the test will be run. Refer to any diagnostic monitors related to the fuel and air induction system that failed prior to the test and confirm that they have run and passed. If possible, drive the vehicle through at least two drive cycles and confirm that the MIL lamp is not illuminated. During test driving, the vehicle should be operated in the same manner as when the fault occurred before repairs were made. Check OBD-II vehicles for pending codes that may have failed once but have not yet caused the MIL to illuminate.

E. Emission Control Systems Diagnosis (8 Questions)

1. Verify customer concern; determine if the concern is the result of a malfunction or normal system operation.

Before diagnosis can begin, the customer's concern must be verified. Knowledge of vehicle operation and of on-board diagnostics will help the technician determine if the concern is normal system operation or the result of a malfunction. Refer to service information and research the normal operation of the vehicle. Accurate descriptions are available to familiarize the technician with normal system operation. Apply this information to the customer's concern to determine if a malfunction exists. An emissions related complaint can range from the MIL being illuminated to the vehicle's failing an emission test. An illuminated MIL can easily be verified. To verify a failed emissions test, the technician will have to run the test to verify the failure and identify the area of failure. The test could include a tailpipe test or a simple monitor readiness status check. The technician can then base the diagnosis around that failure. The conditions under which the failure occurred should be noted and kept track of, as they will aid in verifying the effectiveness of the repair. Conditions that should be noted can include the amount of time the vehicle has been driven, engine temperature, engine load, RPM, vehicle speed, etc. Identify any abnormal scan tool data readings that may help in diagnosis of the vehicle.

2. Inspect and test for missing, modified, inoperative, or tampered components.

Some of the emission-control systems and devices most susceptible to modification and tampering are:

- *Air-injection systems*—The air pump may be removed entirely, or only the belt may be removed. Air-injection nozzles are sometimes plugged along with removal of the pump, hoses, and belt. Small pipe plugs screwed into exhaust manifold ports and crankshaft pulleys with no belts attached are frequent signs of missing air-injection systems.
- *EGR systems*—EGR valves are sometimes removed and their ports sealed with blank flanges or plugs. Some EGR valves can be disabled by hitting the top of the valve with a heavy hammer to increase internal spring tension and keep the diaphragm from opening the valve. Vacuum lines may be plugged with small bolts and ball bearings. Vacuum solenoids may be disconnected or removed. EGR tampering not only increases NO_x emissions, but it can also lead to serious engine detonation and damage because EGR is the principal means used on late-model engines to control pinging.
- *Exhaust systems*—Catalytic converters are commonly cut off either in an attempt to increase engine power or to remove a failed converter. Removing a catalytic converter or replacing it with an unapproved device can cause increased emissions and possibly DTCs.

Many emission inspection programs require functional (operational) tests of air injection and EGR systems. These functional tests often reveal tampering that simple inspection cannot detect.

3. Locate relevant service information.

Service information for emission-control systems originates with the vehicle manufacturers, as does all other service information previously discussed. The kinds and sources of information are related to those for fuel, ignition, and general control system diagnosis and repair.

Two of the most important kinds of service information for emission-control testing and service are a component locator manual and an emission-application manual. An emission-application manual tells what kinds of emission-control systems and devices a vehicle was originally equipped with. A component locator manual tells where they are installed and what they look like.

Additionally, vacuum hose diagrams are essential for emission service. That is why the underhood decal contains the basic vacuum hose routing diagram for the emission devices. Remember, however, that the diagram on the decal is not the complete vacuum system for the vehicle. That is why an additional vacuum diagram manual is also necessary.

4. Research system operation using technical information to determine diagnostic procedure.

Technical information related to emission-control systems will use application tables listing systems used, theory of operation, system functional tests, and component replacement procedures. With this information and a clear description of the problem, the technician can proceed with a systematic approach to diagnosis. While testing emission-control subsystems is a part of any repair where the complaint was a failed emission test, it should be understood that malfunctioning emission-control systems can also cause a host of other driveability problems. Late-model emission controls are a part of the power plant design and not just tagged-on afterthoughts. A problem with the EGR system can cause not only increased NO_x emissions but also engine misfire or hesitation, rough idle and stalling, and engine damage from detonation. It should be clear that proper operation of emission-control systems is essential.

5. Use appropriate diagnostic procedures based on available vehicle data and service information; determine if available information is adequate to proceed with effective diagnosis.

The analytical methods and thought processes for effectively diagnosing emission-control systems is similar to other systems already discussed. An effective diagnosis is very difficult without proper test procedures and a clear understanding of system operation. Manufacturer-specific test procedures must be followed due to the large design differences among the many carmakers. Attempting to diagnose EGR or EVAP systems without vehicle-specific test procedures is difficult not only because these systems vary between carmakers, but also because there may be differences among the same manufacturer's carlines.

6. Determine the relative importance of displayed scan tool data.

All of the data parameters described for ignition, fuel, air intake, and general computer system operation can affect vehicle emissions and be symptoms of emission problems. The following specific data parameters, however, are common ones for specific emission subsystems:

- *Air-injection valve operation*—Most control systems that transmit serial data include switch parameters (on-off or yes-no) that indicate the operation of air switching and air diverter valves. These data items will principally display if air is routed to the exhaust manifold for engine warm-up, directed downstream to the catalyst, or diverted to the atmosphere.
- *EGR system operation*—Many different data parameters indicate EGR valve operation. Some systems provide switch signals (on-off or yes-no) for EGR solenoids. Some systems transmit duty-cycle readings for pulse-width-modulated EGR valves. Others provide position sensor readings as a percentage of valve opening.
- *EVAP system control data*—Many different data parameters indicate operation of EVAP system components. Most often, switch parameters are used to indicate operation of purge and vent solenoids.
- *Catalytic converter monitoring*—A catalytic converter that is not performing up to standards can be verified by watching the upstream and downstream O_2 sensor readings. A downstream O_2 sensor should switch at a much slower rate than the upstream. This indicates that the converter is functioning properly. A downstream O_2 sensor that mimics the switching of the upstream sensor indicates that the catalytic converter is not performing as it should and is not further reducing the pollution levels of the vehicle.

Vehicle manufacturers' system descriptions and test procedures are particularly important for determining normal and abnormal indications for many of these emission data parameters.

7. Differentiate between emission control system's mechanical and electrical/electronic problems.

Because so many emission-control mechanical systems are electronically controlled, it is very important to differentiate between electrical or mechanical problems. A trouble code may be set for an out-of-range EGR valve position sensor, leading the technician to believe the vehicle has an electrical problem when something as simple as a piece of carbon may be holding the valve open and causing a higher-than-normal sensor signal. Computer scan data may indicate that the air-injection system is delivering air to the catalytic converter, but a melted air switching valve can cause the air to be sent to the exhaust manifold instead. A failed exhaust check valve that allowed exhaust gas to flow back into the switching valve may be the root cause of the problem.

> *Note:* Tasks 8 through 12 refer to the following emission control subsystems: Positive crankcase ventilation, ignition timing control, idle and deceleration speed control, exhaust gas recirculation, catalytic converter system, secondary air injection system, intake air temperature control, early fuel evaporation control, and evaporative emission control (including ORVR).

8. Differentiate between driveability or emissions problems caused by failures in emission control systems and other engine management systems.

> *Note:* Tasks E.8 through E.12 refer to the following emission control subsystems: Positive crankcase ventilation, ignition timing control, idle and deceleration speed control, exhaust gas recirculation, catalytic converter system, secondary air injection system, intake air temperature control, early fuel evaporation control, and evaporative emission control which includes onboard refueling vapor recovery. (including ORVR) and engine off natural vacuum (EONV).

The most common reason for needing to diagnose an emission-control subsystem is either a failed emissions test or a DTC indicating a fault with an emissions related device. Other problems that relate to emissions system components could be oil leaks due to malfunctions in the PCV system, or the customer not being able to fill the gas tank with fuel, which points to a problem with the EVAP vent. Emission system faults can also lead to a number of drivability complaints, depending upon which system is malfunctioning.

9. Perform functional tests on emission control subsystems; determine needed repairs.

> *Note:* Tasks E.8 through E.12 refer to the following emission control subsystems: Positive crankcase ventilation, ignition timing control, idle and deceleration speed control, exhaust gas recirculation, catalytic converter system, secondary air injection system, intake air temperature control, early fuel evaporation control, and evaporative emission control which includes onboard refueling vapor recovery. (including ORVR) and engine off natural vacuum (EONV).

Functional testing of emission-control subsystems requires the technician to test all aspects of a particular system. Functional testing of a vacuum operated EGR system includes manually lifting the EGR valve diaphragm to see if the engine stalls (indicating that EGR gases flowed into the intake) and the EGR passages are clear, checking for the presence of vacuum at the valve on systems operated by vacuum, testing the vacuum controls for proper operation and the hoses for leaks or obstructions, and monitoring any feedback sensors for correct signals. Electronic systems can be tested by viewing the EGR position, pressure, or temperature sensor voltages with the scan tool. The system can also be tested by electronically actuating the valve and verifying its operation either by monitoring engine response while the valve is open or by physically watching the opening and closing of the valve. With the valve open and the engine running, the idle should become rough and unstable if the passages are clear.

A secondary air-injection system must be tested to insure that air is delivered to the exhaust manifolds in open-loop operation, that the air switches to the catalytic converter in closed loop, and that air is diverted to the atmosphere during deceleration conditions. A test with an infrared gas analyzer will confirm the presence of oxygen in the exhaust stream, which will thereby verify that no restrictions are present and that the pump output is sufficient. A catalytic converter should be tested for a broken monolith by tapping the converter shell with a rubber mallet; rattling indicates a broken monolith and converter replacement.

An exhaust back pressure test should be performed to test for restrictions. A delta temperature test with a temperature probe can be used to see if the converter will light off. Cranking CO_2 tests, snap throttle oxygen storage tests, or propane conversion tests can be done to test converter efficiency.

10. Determine the effect on exhaust emissions caused by a failure of an emission control component or subsystem.

> *Note:* Tasks E.8 through E.12 refer to the following emission control subsystems: Positive crankcase ventilation, ignition timing control, idle and deceleration speed control, exhaust gas recirculation, catalytic converter system, secondary air injection system, intake air temperature control, early fuel evaporation control, and evaporative emission control which includes onboard refueling vapor recovery. (including ORVR) and engine off natural vacuum (EONV).

This task requires that the technician have a good understanding of the theory of operation for each emission subsystem and its job in controlling tailpipe emissions and also what effect the system can have on engine performance when there is a malfunction. For instance, the EGR system primarily controls NO_x emissions by lowering combustion chamber temperatures below 2500°F (1371°C), but if the EGR valve sticks open, exhaust gas will displace oxygen, which can cause a density misfire in the engine. A misfire such as this will raise hydrocarbons. A failure in the secondary air-injection system may cause air to be continuously delivered upstream ahead of the oxygen sensor. This problem will fool the computer into reading the fuel control as too lean, and the computer will respond by sending a constant rich command. The increased fuel delivery will cause a rise in CO emissions and a possible emission test failure. A catalytic converter failure can cause an increase in all the harmful tailpipe emissions, because modern three-way catalysts control HC, CO and NO_x emissions.

All other emission subsystems must be checked and in good working order for the catalytic converter to work efficiently. A catalytic converter cannot reduce hydrocarbons to an acceptable level if misfire is present in the engine, and the EGR system must reduce formation of NO_x in the combustion chamber so that the converter can bring the level low enough to pass enhanced emission inspections.

11. Use exhaust gas analyzer readings to diagnose the failure of an emission control component or subsystem.

> *Note:* Tasks E.8 through E.12 refer to the following emission control subsystems: Positive crankcase ventilation, ignition timing control, idle and deceleration speed control, exhaust gas recirculation, catalytic converter system, secondary air injection system, intake air temperature control, early fuel evaporation control, and evaporative emission control which includes onboard refueling vapor recovery. (including ORVR) and engine off natural vacuum (EONV).

Using a four- or five-gas analyzer, the technician will perform tests to baseline a vehicle with an emission problem and verify any repairs that are completed. A four-gas analyzer will be useful for problems causing an increase in HC or CO emissions, but a portable five-gas analyzer is needed for NO_x emission problems. Because NO_x emissions are primarily produced under road load conditions, a stationary engine analyzer with NO_x measurement capabilities is not very effective for diagnosing NO_x emission problems. Using tables F1 and F2 located in the next section of this overview, a technician can develop a systematic approach to troubleshooting emission component or subsystem problems based on the readings gathered with a gas analyzer.

12. Diagnose hot or cold no-starting, hard starting, stalling, engine misfire, spark knock, poor driveability, incorrect idle speed, poor idle, flooding, hesitation, backfire, surging, power loss, reduced fuel economy, illuminated MIL, and emission problems caused by a failure of emission control components or subsystems.

> *Note:* Tasks E.8 through E.12 refer to the following emission control subsystems: Positive crankcase ventilation, ignition timing control, idle and deceleration speed control, exhaust gas recirculation, catalytic converter system, secondary air injection system, intake air temperature control, early fuel evaporation control, and evaporative emission control which includes onboard refueling vapor recovery. (including ORVR) and engine off natural vacuum (EONV).

Refer to Tables F1 and F2 at the end of this Overview section. These tables summarize how HC, CO, and NO_x emissions relate to various engine operating conditions and common problems. They also summarize how CO_2, and O_2 present in the exhaust will change under different circumstances.

When any of the gas measurements are out of the normal range, understanding the relationships of these exhaust gases will direct the technician to general area tests of the emission-control components and subsystems listed above.

13. Determine root cause of failures.

14. Determine root cause of multiple component failures.

15. Determine root cause of repeated component failures.

Tasks 13, 14, and 15 are related to each other and to similar tasks listed for other subsystems. Replacing a failed catalytic converter without testing for other engine or emission system problems can lead to an expensive comeback. A spark plug wire arcing to the cylinder head can raise HC emissions and damage a catalytic converter. Newer vehicles are equipped with OBD-II certified emissions systems that monitor for engine misfire to prevent this type of problem. On pre-OBD-II systems it is the job of the technician to determine why a component failed. The arcing spark plug wire mentioned above could be the root cause of the catalytic converter failure as well as the source of repeated catalytic converter failures.

16. Verify effectiveness of repairs.

After repairs have been made, steps should be taken to verify the effectiveness of the repair. Re-perform any tests that were made during diagnosis to verify that the system passes. Refer to any scan tool data, such as monitor readiness status, that will ensure proper system operation. If the condition set a DTC, refer to the enabling criteria for that particular DTC and operate the vehicle under those specific conditions so that the test will be run. Refer to any diagnostic monitors related to the emissions system that failed prior to the test and confirm that they have run and passed. If possible, drive the vehicle through at least two drive cycles and confirm that the MIL lamp is not illuminated. During test driving, the vehicle should be operated in the same manner as when the fault occurred before repairs were made. Check OBD-II vehicles for pending codes that may have failed once but have not yet caused the MIL to illuminate.

F. I/M Failure Diagnosis (6 Questions)

1. Verify customer concern; determine if the concern is the result of a malfunction or normal system operation.

A typical inspection and maintenance (I/M) failure complaint will be accompanied by papers from the testing site indicating the type of failure. This failure can be due to the MIL being illuminated or the vehicle failing an emissions tailpipe test. Testing procedures can vary from one area to another and can also change from one vehicle to another. Knowledge of vehicle operation, on-board diagnostics, and the I/M testing procedure used will help the technician determine the malfunction.

2. Inspect and test for missing, modified, inoperative, or tampered components.

Changing, bypassing, or tampering with electronic engine and emissions control components violates federal law. Part of troubleshooting an I/M failure is to inspect it for missing, damaged, or tampered components, the same as with any other vehicle system. Refer to vehicle-specific information on emission system components prior to inspecting the system.

The first step when diagnosing I/M failures should be a thorough visual inspection.

Regarding air-injection systems: Most newer vehicles no longer have an air pump to check. Some vehicles with V-8 engines are equipped with an electric air pump controlled by the ECM. This air pump operates only during cold start up and WOT. Refer to Emission-Control Systems Diagnosis.

3. Locate relevant service information.

Manufacturers are now providing specific information on properly performing an I/M test on their brand of vehicle. Though these procedures are usually generalized and can be used for any brand, this information should be located and reviewed for special procedures. This information is generally available in service manuals, through online sources, and from aftermarket retailers. Before performing any I/M tests, be sure to refer to any TSBs that may have been issued relating to the test.

4. Evaluate I/M test emission readings to assist in emission failure diagnosis and repair.

Determining which operational mode the vehicle was in during the failure is a very important step in diagnosis. Knowing the operational mode narrows the possible causes of the failure. During certain operational modes, the processor does not look at some input devices, and at other times the processor is in a fixed or modified fuel and timing delivery mode. The various operational modes will change from vehicle to vehicle and will require vehicle-specific information.

Using the readings obtained from an I/M test will provide the technician with the emissions failure, how much it failed by, and which mode it failed in. Using this information will help the technician decide the proper diagnostic and repair procedures.

For example: A vehicle fails the I/M test for high NO_x during acceleration only. Knowing how the NO_x is reduced and by which component would lead the technician to the EGR valve operation. If the EGR is controlled by the ECM, knowing the operational mode during failure may help pinpoint the problem. If the vehicle failed NO_x at all times, the number of systems and components to be checked would increase.

5. Evaluate HC, CO, NO_x, CO_2, and O_2 gas readings; determine the failure relationships.

Tables F1 and F2 (found at the end of this Overview section) summarize how HC, CO, NO_x, CO_2 and O_2 emissions relate to various engine operating conditions and common problems. Refer to these tables to review the failure relationships. The Vehicle Inspection Report (VIR)—provided to the vehicle owner if the vehicle fails an enhanced IM240 emission test—is extremely important because it displays the vehicle's emissions output on a graph so that the technician can see how the operating conditions affected the emissions. The vehicle drive trace is superimposed on each exhaust gas trace so that the technician can compare emissions produced to how the vehicle was being driven, such as idling, accelerating, decelerating, or steady cruise conditions. Elevated emissions outputs under certain driving conditions can point the technician in the proper direction for the diagnosis of many problems. For instance, a sticking EGR valve can cause HC spikes during decel conditions; a saturated charcoal canister may cause elevated CO emissions during steady state cruise when purge is commanded on; and a weak catalytic converter may produce a drive trace showing elevated HC, CO and NO_x during the high-speed portion of the drive trace when exhaust flow is highest.

6. Use test instruments to observe, recognize, and interpret electrical/electronic signals.

Electrical and electronic test equipment includes ignition oscilloscopes, laboratory oscilloscopes (lab scopes), digital volt-ohm-ammeters (DVOM), and scan tools.

Ignition scope tests can reveal ignition problems; but just as importantly, abnormalities in ignition voltage waveforms can indicate fuel system and mechanical problems.

Labscope testing has an advantage over basic voltmeter testing, because an oscilloscope displays voltage changes over a period of time. The two-dimensional waveform, or trace, can reveal irregularities (glitches) that a voltmeter would not indicate.

Waveform analysis of the oxygen sensor with a labscope is one of the most important tests to perform when diagnosing emission test failures. An oxygen sensor waveform can provide information such as whether or not misfire is present during driving conditions and whether or not the computer is in proper fuel control. A proper oxygen sensor waveform provides confirmation of normal fuel control. Normal oxygen sensor switching frequency is critical to proper catalytic converter operation. A properly-operating fuel control system on a fuel-injected engine will show an oxygen-sensor switching rate between 0.5 to 5 Hz at 2500 rpm. Oxygen-sensor switching frequency will decrease at low RPM or idle conditions and should always be checked at 2000 to 2500 rpm when comparing to specifications.

When testing an oxygen sensor with a labscope, it is easiest to calculate switching frequency by setting the scope time-base to display ten seconds of time across the scope screen. Using this setting, if the oxygen sensor switches 10 times during one screen sweep, the frequency is 1 Hz. If 18 switches are shown, the frequency is 1.8 Hz. Slow or lazy oxygen sensors throw off fuel control and reduce converter efficiency. This problem may not be apparent if the oxygen sensor is tested with a voltmeter, but it is clearly visible when tested with a labscope.

7. Analyze HC, CO, NO_x, CO_2, and O_2 readings; determine diagnostic test sequence.

Tables F1 and F2 (found at the end of this Overview section) summarize how HC, CO, NO_x, CO_2, and O_2 emissions relate to various engine operating conditions and common problems. Refer to these tables to review the failure relationships.

When a particular combination of exhaust analyzer readings and emissions that exceed test limits is understood, the most appropriate test and repair sequences can be chosen based on the most probable causes listed in Tables F1 and F2. These procedures most often will be the same ones reviewed in previous sections of this overview.

8. Diagnose the cause of no-load I/M test HC emission failures.

Hydrocarbons (HC) are unburned fuel molecules. No-load I/M test HC emission failures indicate a problem with combustion efficiency. Excessive HC present in the exhaust can be caused by anything that affects the combustion process. Items to be considered for causing high HC include cylinder compression and airflow, ignition operation and timing, air/fuel mixtures, and possible engine modifications or tampering. A failed catalytic converter can also cause HC emissions to be excessive. Always remember: After diagnosing a failed converter, locate the root cause of the failure. High HC output from the engine, whether from unburned fuel, excessive blowby, or oil consumption, can damage the catalytic converter and cause it to function poorly.

9. Diagnose the cause of no-load I/M test CO emission failures.

Carbon monoxide present in the exhaust stream indicates there was excessive fuel present for the amount of air in the cylinder during combustion. This is primarily a rich mixture indicator. Review Table F1 headings "Very Rich—Below 10:1 at all speeds" and "Rich—10:1 to 12:1 at low speed only" for a listing of common causes for excessive CO emissions. Most no-load I/M test CO emission failures can be traced to one of the causes listed there.

10. Diagnose the cause of loaded-mode I/M test HC emission failures.

Causes for HC failures during loaded-mode I/M testing are the same as for no-load testing, but the engine operating conditions are different, and therefore diagnostic strategies must be changed. Problems that may not show up under static engine testing can become an issue when the engine is operated under load. An arcing spark plug wire or lean injector may not cause a misfire until the engine is loaded; these types of problems need to be identified. Oxygen sensor waveform analysis is one good method for determining if this type of problem exists. Portable gas analyzers also become much more useful in tracking down emission problems that require testing in a loaded condition.

11. Diagnose the cause of loaded-mode I/M test CO emission failures.

Strategies for diagnosing the cause of loaded-mode I/M test CO emission failures are similar to those discussed in Task F.8. For instance, many vehicles may not command charcoal canister purge until the computer sees a vehicle speed input. Exhaust gas testing in the service bay may show normal CO readings, leading the technician to believe the engine is operating properly. A portable gas analyzer can show high CO output from the canister purging, because measurements are taken under normal driving conditions. The technician needs to understand how the various systems operate and how different driving conditions can cause problems to occur.

12. Diagnose the cause of loaded-mode I/M test NO_x emission failures.

Loaded-mode I/M test NO_x emission failures are one of the most challenging emission test failures for technicians to diagnose. Many engine systems can contribute to excessive NO_x emissions.

While EGR is the single most important system in controlling NO_x, problems in the cooling system that raise engine-operating temperature, over-advanced ignition timing, or a failed catalytic converter will all contribute to increased NO_x production. Even correcting a different emission problem can create an increase in NO_x levels. Carbon deposits formed in the combustion chamber from running an engine with an excessively rich air/fuel mixture can increase compression pressure. Repairing the rich running condition will lean the air/fuel mixture, but the remaining high compression from the carbon deposits can cause NO_x levels to go up.

Performing a de-carbon treatment to the engine prior to an emission re-test is recommended in this case. This condition has caused many emission re-test failures. Baselining a vehicle with a five-gas analyzer at the beginning of diagnostics and repeating the test after repairs to verify their effectiveness is necessary to prevent the vehicle from failing a re-test.

13. Evaluate the MIL operation for on-board diagnostic I/M testing.

Generally when a fault has occurred that causes emissions levels to increase 1.5 times the federal limit, a DTC will set and the MIL will illuminate. Certain conditions, such as a catalyst damaging fault (a severe misfire, for example), may cause the MIL to flash. To test the operation of the MIL, turn the key on and confirm that the MIL illuminates for approximately two seconds. This verifies that the MIL lamp and circuit are operational. Refer to vehicle specific information for a detailed description of MIL operation.

14. Evaluate monitor readiness status for on-board diagnostic I/M testing

OBD-II monitors are in charge of testing certain areas of engine operation and reporting failures that can lead to high emission output. Typical OBD-II monitors include: Misfire, Fuel Trim, Comprehensive Components, Catalyst, Exhaust Gas Recirculation, Evaporative Emissions, Oxygen (O_2) Sensor, Secondary Air, Heated Catalyst, O_2 Sensor Heater, and Positive Crankcase Ventilation. The status for these monitors can be viewed with the scan tool. The monitor can be displayed in several ways. A monitor status of "Ready" or "Complete" indicates that the monitor has run all tests necessary to verify the proper operation of the system. A monitor status of "Not Ready," "Not Complete," or "Incomplete" may indicate that the monitor has not yet met all of the enabling criteria to run the monitor and that all tests have not been performed. A "Not Ready" status can also indicate that the monitor has failed and DTCs have been set.

15. Diagnose communication failures with the vehicle during on-board diagnostic I/M testing.

To check the status of the OBD-II monitors or obtain DTCs, the scan tool must be able to communicate with the ECM. A communication failure could be due to simple faults such as blown fuses resulting in no power at the

DLC, improper ground circuit of the DLC, or pushed in or spread terminals at the DLC. The fault could also be the result of a problem in the data bus network. Make all of the simple checks first before trying to diagnose a problem with the network. If the vehicle runs and was driven to the shop for the I/M test, it is not likely that there would be a major problem with the vehicle's network.

16. Perform functional I/M tests (including fuel cap test).

I/M tests can be performed in many different ways depending upon the city or state in which the vehicle is being tested. I/M programs are generally divided into two programs: *Basic* I/M testing and *Enhanced* I/M testing.

A basic I/M testing program may consist of a one- and/or two-speed idle test that measures tailpipe emissions at idle speed and fast idle speed around 2500 rpm. Visual tampering checks may also be made to certain areas that include, but are not limited to, the catalytic converter, fuel tank inlet restrictor, EGR system, EVAP system (including a pressure check of the fuel cap), air injection system, and PCV system. The basic test can also include a "plug in" test, which checks the status of the OBD-II monitors. Any monitors that have not run and passed or an illuminated MIL will cause the vehicle to fail the test. Basic I/M tests may be performed by repair facilities rather than a government-operated facility.

An enhanced I/M testing program can include the same tests that the basic program uses, but the vehicle will also have to perform a loaded mode test on a dyno. While on the dyno, different driving conditions can be simulated while the exhaust is tested for HCs, CO, CO_2, O_2, and NO_x levels. The most commonly used test is the I/M 240. This test runs the vehicle on the dyno for 240 seconds while simulating certain driving conditions and noting emissions levels during these conditions. During this time the EVAP system and fuel cap are thoroughly tested.

17. Verify effectiveness of repairs.

Upon completion of the repairs, the vehicle should be re-tested and the results compared to the first test that the vehicle failed. Improvements should be seen in the test results; the bigger the improvement the more effective the repair.

Some portable gas analyzers allow before-and-after emission snapshots to be taken, similar to snapshots that are done with a scan tool. The analyzer will compare two snapshots and display the actual change in percentage of the emissions measured. Using this feature, the technician can determine if the emissions improved or got worse and by how much. This feature may also be available by connecting the gas analyzer to a PC with dedicated software created by the equipment manufacturer. In either case, a direct means of measuring repair effectiveness is possible.

Table F1

AIR/FUEL RATIO	EXHAUST EMISSIONS				
	Engine Speed	HC	CO	CO_2	O_2
Very Rich—Below 10:1 at all speeds	Idle	250 ppm	3%	7 to 9%	0.2%
	Off Idle	275 ppm	3%	7 to 9%	0.2%
	Cruise	300 ppm	3%	7 to 9%	0.2%
Other Symptoms	Black smoke or sulfur odor, poor fuel economy, surge or hesitation, stalling, rough or "lumpy" idle, engine not warming to operating temperature, continuous open-loop operation				
Possible Causes	• High MAP sensor voltage (vacuum leak or electrical fault) • Leaking fuel injectors • High fuel pressure • Thermostat stuck open or engine otherwise continuously operating at very low temperature				

EXHAUST EMISSIONS

AIR/FUEL RATIO	Engine Speed	HC	CO	CO_2	O_2
Rich—10:1 to 12:1 at low speed only	Idle	150 ppm	1.5%	7 to 9%	0.5%
	Off Idle	150 ppm	1.5%	7 to 9%	0.5%
	Cruise	100 ppm	1.0%	11 to 13%	1.0%

Other Symptoms

Poor fuel economy, surge or hesitation, black smoke and soot-fouled spark plugs, rough idle, vapor canister saturated with fuel or purge valve bad

Possible Causes

- High MAP sensor voltage (vacuum leak or electrical fault)
- Leaking fuel injectors
- Engine oil diluted with gasoline
- Excessive crankcase blowby
- High fuel pressure
- Thermostat stuck open or engine otherwise continuously operating at very low temperature

EXHAUST EMISSIONS

AIR/FUEL RATIO	Engine Speed	HC	CO	CO_2	O_2
Very Lean—Above 16:1 at all speeds	Idle	200 ppm	0.5%	7 to 9%	4 to 5%
	Off Idle	205 ppm	0.5%	7 to 9%	4 to 5%
	Cruise	250 ppm	1.0%	7 to 9%	4 to 5%

Other Symptoms

Rough idle, high-speed misfire, overheating, surging, hesitation, detonation at cruising speeds

Possible Causes

- Intermittent ignition problems causing misfire
- Restricted fuel injectors
- Low fuel pressure
- Vacuum leak
- Poor cylinder sealing (low compression)
- Improper ignition timing
- Thermostat stuck closed or engine otherwise operating at very high temperature

EXHAUST EMISSIONS

AIR/FUEL RATIO	Engine Speed	HC	CO	CO_2	O_2
Lean—Above 16:1 at high speed	Idle	100 ppm	2.5%	7 to 9%	2 to 3%
	Off Idle	80 ppm	1.0%	7 to 9%	2 to 3%
	Cruise	50 ppm	0.8%	7 to 9%	2 to 3%

Other Symptoms

Rough idle, misfire, surging, hesitation

Possible Causes

- Intermittent ignition problems causing misfire
- Restricted fuel injectors
- Low fuel pressure
- Vacuum leak
- Heated air intake stuck in cold-air position

EXHAUST EMISSIONS

AIR/FUEL RATIO	Engine Speed	HC	CO	CO$_2$	O$_2$
Normal—13:1 to 15:1 but engine not fully warmed up	Idle	100 ppm	0.3%	10 to 12%	2.5%
	Off Idle	80 ppm	0.3%	10 to 12%	2.5%
	Cruise	50 ppm	0.3%	10 to 12%	2.5%

Other Symptoms	Cold-engine emission test failure, catalytic converter not warmed up

Table F2

ENGINE PROBLEM	EXHAUST EMISSIONS				
	HC	CO	CO$_2$	O$_2$	NO$_x$
Rich Mixture	Moderate Increase	Large Increase	Some Decrease	Some Decrease	Moderate Decrease
Lean Mixture	Moderate Increase	Large Decrease	Some Decrease	Some Increase	Moderate Increase
Very Lean Mixture	Large Increase	Large Decrease	Some Decrease	Large Increase	Large Increase
Ignition Misfire	Large Increase	Some Decrease	Some Decrease	Moderate Increase	Moderate Decrease
Advanced Timing	Some Increase	No Change or Slight Decrease	No Change	No Change	Large Increase
Retarded Timing	Some Decrease	No Change or Slight Increase	No Change	No Change	Large Decrease
Very Retarded Timing	Some Increase	No Change	Moderate Decrease	No Change	Some Increase
Low Compression	Moderate Increase	Some Decrease	Some Decrease	Some Increase	Moderate Decrease
Exhaust Leak	Some Decrease	Some Decrease	Some Decrease	Some Increase	No Change
Worn (Flat) Cam Lobes	No Change or Some Decrease	Some Decrease	Some Decrease	No Change or Some Decrease	No Change or Some Decrease
General Engine Wear	Some Increase	Some Increase	Some Decrease	Some Decrease	No Change or Slight Decrease
Air Injection Failure	Some Increase	Large Increase	Moderate Decrease	Moderate Decrease	No Change
EGR Leaking	Some Increase	No Change	No Change or Some Decrease	No Change	No Change or Some Decrease

NORMAL EMISSION CONTROL CONDITION	HC	CO	CO$_2$	O$_2$	NO$_x$
EGR Operating Normally	No Change	No Change	Some Decrease	No Change	Large Decrease
Air Injection Operating Normally	Large Decrease	Large Decrease	Moderate Decrease	Large Increase	No Change

COMPOSITE VEHICLE

1. You will notice as you read through the Task List that the job skills identified concentrate on the ability to diagnose, rather than repair. The panel of experts who developed the L1 test have identified three important general characteristics of drivability diagnosis.

2. Data is obtained from the vehicle using a variety of test instruments and is compared to known values obtained from the service manuals.

3. A good technician can draw valid conclusions from the relationship between published data and what he understands of the vehicles fuel, ignition, and emission-control systems.

The ASE panel of experts has developed a "Composite Vehicle" engine control system, which is described in detail in the "Composite Vehicle Preparation/Registration Booklet" you'll receive when you register for the test. The information is included here so that you can begin the familiarization process now.

The composite vehicle uses a "mass airflow" fuel-injection system design used by many domestic and import manufacturers. The system uses sensor, actuators, and control strategies that you should be familiar with from your shop experience. When you answer questions based on the Composite Vehicle, you will be simulating your real-world experience of using reference materials and test instruments to diagnose problems based on your knowledge of a particular engine-management system.

The following Composite Vehicle Information has been provided by the National Institute for Automotive Service Excellence (ASE). Delmar, Cengage Learning would like to thank ASE for providing this content for use in this study guide.

An electronic version of the Composite Vehicle reference booklet is also available at www.ase.com.

This booklet is intended only for reference when preparing for and taking the ASE Advanced Engine Performance (L1) Test. The automotive composite vehicle control system is based on designs common to many engine and vehicle manufacturers, but is not identical to any actual production engine or vehicle.

AUTOMOTIVE COMPOSITE VEHICLE INFORMATION

INTRODUCTION

This ASE Composite Vehicle Type 4 was conceived and built by technical committees of industry experts to accommodate high level diagnostic questions on the L1 test. While some aspects of this Composite Vehicle may appear similar to vehicles from a number of manufacturers, it is important to understand this vehicle is a unique design and is NOT intended to represent any specific make or model. This reference document should be used when answering questions identified as Composite Vehicle questions.

Note: All testing is performed at sea level unless otherwise indicated. The reference materials and questions for this test use terms and acronyms that are consistent with SAE standards J1930 and J2012.

POWERTRAIN

Engine

- Generic, four-stroke, V6 design.
- Equipped with four chain-driven overhead camshafts, 24 valves, hydraulic valve lifters, variable intake camshaft timing, and variable intake valve lift.

Transmission

- 6-speed, automatic transaxle with overdrive.
- Controlled by a transmission control module (TCM).

- 3 planetary gear sets, 5 clutch packs, and a single one-way clutch.
- 6 forward gears and 1 reverse gear.
- A torque converter transmits power from the engine to the transmission and is capable of lock-up in 3rd, 4th, 5th, and 6th gears.
- Contains an electronic pressure control (EPC) solenoid, 5 shift solenoids, and a torque converter clutch solenoid.

CONTROL MODULES

Engine Control Module (ECM)

- Calculates ignition and fuel requirements, controls engine actuators and provides inputs to other modules to provide the desired driveability, fuel economy, and emissions control.
- Receives data input from other control modules and sensors.
- Controls the vehicle's charging system.
- Receives power from the battery and ignition switch and provides a regulated 5-volt supply for most of the engine sensors.
- Engine control features include coil-on-plug ignition, mass airflow, sequential port fuel injection, variable valve timing, variable valve lift, electronic throttle actuator control (TAC), air/fuel ratio sensors, a data communications bus, a vehicle anti-theft immobilizer system, a natural vacuum leak detection EVAP system and an on-board refueling vapor recovery (ORVR) system.
- The control system software and OBD diagnostic procedures stored in the ECM can be updated using factory supplied calibration files and PC-based interface software, along with a scan tool or a reprogramming device that connects the PC to the vehicle's data link connector (DLC).
- Contains a 120 Ω terminating resistor for the data bus.

Fuel Pump Control Module (FPCM)

- Communicates with the ECM over a Local Area Network (LAN).
- ECM provides a 5-volt enable signal to the FPCM to enable fuel pump operation:
  - for two seconds with the ignition switch in the RUN position.
  - when the ignition switch is in the START position.
  - when the engine speed (CKP) signal is above 100 rpm.
- FPCM changes the volume of fuel supplied by the fuel pump by varying the duty-cycle of the voltage supplied to the fuel pump.
- LOW fuel pump speed command = fuel pump supply voltage duty-cycled at 50 %.
- HIGH fuel pump speed command = fuel pump supply voltage duty-cycled at 100 %.
- HIGH fuel pump speed is commanded:
  - during key ON/engine OFF prime.
  - with the engine cranking.
  - under high engine load.
  - during operation at low charging system voltage.
- If there is a communication fault on the LAN bus, and the 5-volt enable signal is present at the FPCM, the FPCM will default to HIGH speed fuel pump operation.
- Actual fuel pump duty cycle is monitored by the FPCM and is reported by the FPCM to the ECM via the LAN bus.

ECM	OFF	LOW	HIGH
FPCM Feedback	0 %	50 %	100 %

Transmission Control Module (TCM)

- Provides the correct transmission outputs for desired driveability, fuel economy, and emissions control.
- Receives data input from other control modules and sensors.
- Provides data inputs to other control modules including vehicle speed and gear selection.
- Provides its own regulated 5-volt supply.
- Performs all OBD II transaxle diagnostic routines and stores transaxle DTCs.
- The control system software and OBD diagnostic procedures stored in the TCM can be updated in the same way as the ECM.
- Failures that result in a pending or confirmed DTC related to any of the following components will cause the TCM to default to fail-safe mode: transmission range switch, electronic pressure control (EPC), shift solenoids, turbine shaft speed sensor, and the vehicle speed sensor.
- The TCM will also default to fail-safe mode if it is unable to communicate with the ECM.
- When in fail-safe mode, the TCM commands maximum line pressure and turns off all transmission solenoids. The transmission then defaults to 5th gear and the torque converter clutch will be disabled.

Instrument Cluster Module (ICM)

- Receives data input from other control modules to display engine rpm, vehicle speed, fuel level, and coolant temperature.
- Includes a Malfunction Indicator Lamp (MIL) and an immobilizer indicator.
- If the instrument cluster fails to communicate with the ECM and TCM, the MIL is continuously lit.
- Contains a 120 Ω terminating resistor for the data bus.

Immobilizer Module

- Communicates with the ECM.
- Provides ignition key information.
- See IMMOBILIZER ANTITHEFT SYSTEM on page 51.

SYSTEMS

Electronic Throttle Control System

- The vehicle does not have a mechanical throttle cable, a cruise control throttle actuator, or an idle air control (IAC) valve.
- Throttle opening at all engine speeds and loads is controlled directly by a throttle actuator control (TAC) motor mounted on the throttle body housing.
- Dual accelerator pedal position (APP) sensors provide input from the vehicle operator, while the actual throttle angle is determined using dual throttle position (TP) sensors.
- If one APP sensor or one TP sensor fails, the ECM will turn on the malfunction indicator lamp (MIL) and limit the maximum throttle opening to 35 %.
- If both APP sensors or both TP sensors fail, or a correlation error occurs, the ECM will turn on the MIL and disable the electronic throttle control.
- When disabled by the ECM, the electronic throttle control system will default to limp-in operation:
    - the spring-loaded throttle plate will return to a default position of 15 % throttle opening.
    - the TAC value on the scan tool will indicate 15 %.
    - it will have a fast idle speed of 1400 to 1500 rpm, with no load and all accessories off.
- Normal no load idle range is 850 to 900 rpm at 5 % to 10 % throttle opening.
- No idle relearn procedure is required after component replacement or loss of voltage to the ECM.

Exhaust System

A single exhaust system that is configured using a Y-pipe that connects two front catalysts, a single downstream catalyst, and a muffler.

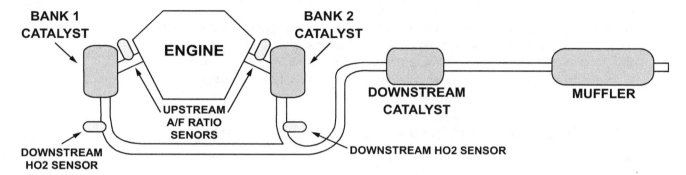

Fuel Delivery System

- Consists of an ECM, a fuel pump control module (FPCM), a fuel pressure sensor (FPS), and a fuel pump assembly.
- Sequential Multiport Fuel Injection (SFI).
- Returnless fuel supply with the electric fuel pump assembly mounted inside the fuel tank.
- Fuel pressure regulator attached to the fuel pump assembly to control fuel pressure.
- The fuel pump control module (FPCM) supplies duty-cycled, feed-side voltage to the fuel pump.
- The fuel pump control module (FPCM) provides feedback to the ECM via a LAN bus.
- Key ON/engine OFF fuel pressure = 58 to 62 psi (400 to 427 kPa).
- Fuel system pressure should be between 58 to 62 psi (400 to 427 kPa) during all operating conditions.

Ignition System

- Distributorless Ignition (EI) with six ignition coils (coil-on-plug).
- Firing Order: 1-2-3-4-5-6
- Cylinders 1, 3, and 5 are on Bank 1; Cylinders 2, 4, and 6 are on Bank 2.
- Ignition timing is not adjustable.
- Crankshaft position (CKP) sensor input is used for base timing calculation.
- ECM controls ignition timing.
- Ignition coil drivers are integrated into the ECM.

Immobilizer Anti-Theft System

- When the ignition switch is turned on, the immobilizer control module sends a challenge signal through the antenna around the ignition switch to the transponder chip in the ignition key. The transponder key responds with an encrypted key code. The immobilizer control module then decodes the key code and compares it to the list of registered keys.
- When the engine is started, the ECM sends a request to the immobilizer control module over the data bus to verify the key validity. If the key is valid, the immobilizer control module responds with a "valid key" message. The ECM continues normal engine operation.
- Once the engine is started with a valid key, the immobilizer system cannot cause engine shutdown.
- If an attempt is made to start the vehicle with an invalid ignition key, the immobilizer control module sends a message over the data bus to the instrument cluster to flash the anti-theft indicator lamp.
- Without a "valid key" message from the immobilizer control module within 2 seconds of engine startup, the ECM will disable the fuel injectors to kill the engine. Cycling the key off and cranking the engine again will result in engine restart and stall.

- The immobilizer control module and ECM each have their own unique internal ID numbers used to encrypt their messages, and are programmed at the factory to recognize each other. If either module is replaced, the scan tool must be used to program the replacement module, using the VIN, the date, and a factory-assigned PIN number.
- Up to eight keys can be registered in the immobilizer control module.
- Each key has its own unique internal key code.
- If only one valid key is available, or if all keys have been lost, the scan tool can be used to delete lost keys and register new keys. This procedure also requires the VIN, the date, and a factory-assigned PIN number.
- The immobilizer control module does not require a key ID relearn if battery voltage is lost.
- Neither the ECM, TCM, nor the immobilizer control module prevent operation of the starter motor for anti-theft purposes.

On-Board Refueling Vapor Recovery (ORVR) EVAP System

- Causes fuel tank vapors to be directed to the EVAP charcoal canister during refueling, so that fuel vapors do not escape into the atmosphere.
- The following components have been added to the traditional EVAP system for ORVR capability: a one inch I.D. fill pipe, a one-way check valve at the bottom of the fill pipe, an ORVR vapor control valve inside the fuel tank, and a 1/2 inch I.D. vent hose from the ORVR vapor control valve to the canister.
- The ORVR vapor control valve has a float that rises to seal the vent hose when the fuel tank is full. It also prevents liquid fuel from reaching the canister and blocks fuel from leaking in the event of a vehicle roll-over.

Fuel Injection System

- Sequential port fuel injection, single injector for each cylinder.
- Fuel injectors are located in the intake manifold ports near the intake valves.
- Fuel injectors are ground-side controlled.

Starting Mode

- When the ignition switch is turned to RUN, the ECM sends a 5-volt enable signal to the FPCM for two seconds to build pressure in the fuel system.
- If an rpm signal is not received by the ECM within two seconds, the 5-volt enable signal to the FPCM is turned OFF.
- After the two second prime, the ECM will maintain the 5-volt enable signal to the FPCM with the ignition switch in the START position, or as long as the engine speed (CKP) is 100 rpm or more.

Clear Flood Mode

- During cranking, when the accelerator pedal is fully depressed (pedal position of 80% or greater) and the engine speed is below 400 rpm, the ECM turns off the fuel injectors.

Run Modes: Open and Closed Loop. Fuel Cut Off

- OPEN LOOP - In open loop, the ECM does not use the air/fuel ratio sensor signals. Instead, it calculates the fuel injector pulse width based on MAF and engine temperature. The system will stay in open loop until all of these conditions are met:
  - ten seconds have elapsed since start up.
  - throttle position is less than 80 %.
- CLOSED LOOP - When the ECM receives valid air/fuel ratio signals and the throttle is open less than 80 %, the system will be in closed loop.
- FUEL CUT OFF MODE - The ECM will turn off the fuel injectors if any of the following are met:
  - vehicle speed reaches 110 mph.
  - engine speed exceeds 6000 rpm while driving.

- engine speed exceeds 3000 rpm in PARK/NEUTRAL.
- vehicle is decelerating with engine speed greater than 1500 rpm, engine temperature is greater than 120° F (49° C), and the throttle is closed (APP less than 10 %).

Absolute Load

- The ECM uses the MAF sensor input and stored engine displacement information versus engine speed to calculate the air charge moving through the engine against a theoretical maximum.
- Values of absolute load correlate with volumetric efficiency at wide open throttle (WOT).
- Displayed as a percentage in scan data.
- Normal absolute load at WOT is 95 %.
- Typical values at normal idle are approximately 15 %.

Variable Valve Lift Control System

- Variable valve lift is used for improved engine efficiency, performance and emissions control.
- The ECM controls variable valve lift in relation to engine rpm. Below 3000 rpm, the VVL system will command motor position to 0 %, base valve lift. Above 3000 rpm, the VVL system will command motor position to 100 % resulting in an additional 4 mm of valve lift.
- Each bank has its own 2-wire motor with position sensors for feedback.
- The VVL position sensors will read 0.50 V at low lift (0 % command) and 4.50 V at high lift (100 % command).
- On each intake camshaft, a DC motor is attached to a rod which operates a fulcrum attached to the rocker arms, effectively changing the rocker arm ratio.
- If there are any faults detected in the VVL system, the VVL will be commanded to the low lift position.
- Anytime the TAC system is in failsafe mode (disabled), the VVL will also be commanded to the lowest lift position.
- The VVL system is not used to control idle speed.

Variable Valve Timing System

- A single timing chain drives the intake and exhaust cams of both banks of the engine.
- Intake camshaft timing is continuously variable using a hydraulic actuator attached to the front end of each intake camshaft. Engine oil flow to each hydraulic actuator is controlled by a camshaft position actuator control solenoid.
- The exhaust camshaft timing is fixed.
- Camshaft timing is determined by the ECM using the crankshaft position (CKP) sensor and camshaft position sensor (CMP 1 and CMP 2) signals.
- At idle, the intake camshafts are fully retarded and valve overlap is zero degrees.
- At higher speeds and loads, the intake camshafts can be advanced up to 40 crankshaft degrees.
- Each intake camshaft has a separate camshaft position sensor, hydraulic actuator, and control solenoid.
- If little or no oil pressure is received by a hydraulic actuator (typically at engine startup, at idle speed, or during a fault condition), it is designed to mechanically default to the fully retarded position (zero valve overlap), and is held in that position by a spring-loaded locking pin.

INPUTS–SENSORS
Accelerator Pedal Position (APP 1 and APP 2) Sensors

- A pair of redundant non-adjustable potentiometers that sense accelerator pedal position.
- Located on the accelerator pedal assembly.
- APP 1 sensor output varies from 0.5 volts (pedal released) to 3.5 volts (pedal fully pressed); increasing voltage with increasing pedal position.
- APP 2 sensor output varies from 1.5 volts (pedal released) to 4.5 volts (pedal fully pressed); increasing voltage with increasing pedal position, offset from the APP 1 sensor signal by 1.0 volt.

- ECM interprets an APP of 80 % or greater as a request for wide open throttle.
- A circuit failure of one APP sensor will set a DTC and the ECM will limit the maximum throttle opening to 35 %.
- A circuit failure of both APP sensors, or a correlation error, will set a DTC and disable the TAC.
- When disabled, the spring-loaded throttle plate will return to the default 15 % position (fast idle).

Accelerator Pedal Position (% applied)	APP 1 Sensor Voltage	APP 2 Sensor Voltage
0	0.50	1.50
5	0.65	1.65
10	0.80	1.80
15	0.95	1.95
20	1.10	2.10
25	1.25	2.25
40	1.70	2.70
50	2.00	3.00
60	2.30	3.30
75	2.75	3.75
80	2.90	3.90
100	3.50	4.50

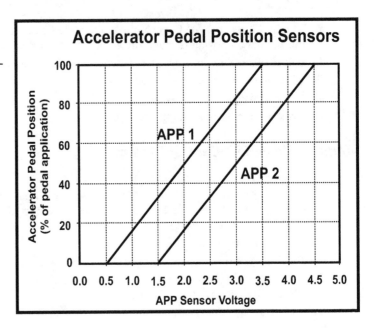

A/C Pressure Sensor

- Three-wire solid-state sensor for A/C system high-side pressure.
- Sensor output varies from 0.25 volts at 25 psi to 4.50 volts at 450 psi.
- Used as input for A/C compressor clutch control, radiator fan control, and idle speed compensation.
- ECM will disable A/C compressor operation if the pressure is below 40 psi or above 420 psi.
- Located on the A/C high-side vapor line.

A/C High Side Pressure (psi)	Sensor Voltage
25	0.25
50	0.50
100	1.00
150	1.50
200	2.00
250	2.50
300	3.00
350	3.50
400	4.00
450	4.50

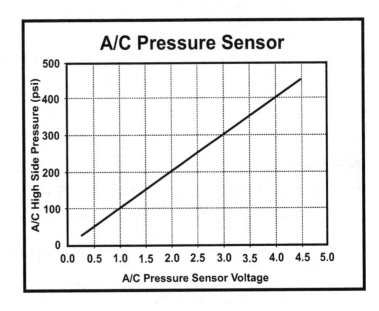

A/C ON/OFF Request Switch

- Normally open (N.O.) switch that closes when A/C compressor operation is requested.
- Status is used by ECM.
- Located in the climate control unit on the instrument panel.

Air/Fuel Ratio Sensors (AFRS 1/1 AND AFRS 2/1)

- Planar-type AFR sensor used by the ECM to measure the air/fuel ratio of the exhaust stream.
- AFRS 1/1 located on the Bank 1 exhaust manifold (cylinders 1, 3, and 5).
- AFRS 2/1 located on the Bank 2 exhaust manifold (cylinders 2, 4, and 6).
- Perfectly balanced air/fuel mixture at 14.7:1 (Lambda 1) = 2.5 volts displayed on the scan tool.
- Lean air/fuel mixture at 20:1 (Lambda 1.36) = 4.3 volts displayed on the scan tool.
- Rich air/fuel mixture at 11:1 (Lambda 0.75) = 1.3 volts displayed on the scan tool.
- ECM monitors polarity and quantity of current to the sensor to determine air/fuel ratio in the exhaust.
- Perfectly balanced air/fuel mixture at 14.7:1 (Lambda 1) = no sensor current produced.
- Rich air/fuel mixture = the sensor produces a negative current be tween zero and −2000 microamps.
- Lean air/fuel mixture = the sensor produces a positive current between zero and +3000 microamps.
- Battery voltage is continuously supplied to the air/fuel ratio sensor heaters when ignition switch is ON.
- The ECM supplies a pulse width modulated ground to the heaters to control the temperature of the sensor. The duty cycle displayed on the scan tool represents the percent of heater current on time.
- The ECM monitors the AFR heater current.
- The normal AFRS heater resistance is 2-6 Ω at 68° F (20° C).

Air/Fuel Ratio	Lambda	Scan Tool Voltage	Current (microamps)
20:1	1.36	4.3	+3000
19:1	1.30	4.0	+2440
18:1	1.23	3.7	+1890
17:1	1.16	3.3	+1330
16:1	1.09	3.0	+780
15:1	1.02	2.7	+220
14.7:1	1.00	2.5	0
14:1	0.96	2.3	−330
13:1	0.89	2.0	−890
12:1	0.82	1.7	−1440
11:1	0.75	1.3	−2000

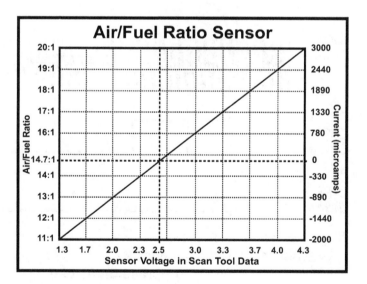

Brake Pedal Position (BPP) Switch

- Normally open (N.O.) switch that closes when the brake pedal is pressed/applied.
- Status is used by TCM.
- Located on the brake pedal.

Camshaft Position Sensors (CMP 1 and CMP 2)

- A pair of three-wire solid state Hall-effect sensors that generate a signal once per intake camshaft revolution.
- Located at the rear of each valve cover, the sensors are triggered by an interrupter on the intake camshafts.
- The leading edge of the bank 1 CMP signal occurs on the cylinder 1 compression stroke, and the leading edge of the bank 2 CMP signal occurs on the cylinder 4 compression stroke.
- When the intake camshafts are fully retarded (zero valve overlap), the CMP signals switch from
- 0 to +5 volts at top dead center compression stroke of cylinders 1 and 4 respectively. When the intake camshafts are fully advanced (maximum valve overlap), the signals switch at 40 crankshaft degrees before top dead center. These signals allow the ECM to determine fuel injector and ignition coil sequence, as well as the actual intake valve timing.
- Loss of one CMP signal will set a DTC, and valve timing defaults to the fully retarded position (zero valve overlap). If neither CMP signal is detected during cranking, the ECM stores a DTC and disables the fuel injectors, resulting in a no-start condition.
- The sensors are not adjustable.
- The diagram at the bottom of this page shows the CKP and CMP sensor signal waveforms with the camshafts at the default (fully retarded) position.

Crankshaft Position (CKP) Sensor

- A magnetic-type sensor that generates 35 pulses for each crankshaft revolution.
- Located on the front engine cover.
- Triggered by a reluctor wheel mounted on the crankshaft, behind the balancer pulley.
- Each tooth is ten crankshaft degrees apart, with one space for a "missing tooth" located at 60 degrees before top dead center of cylinder number 1.
- The diagram at the bottom of this page shows the CKP sensor signal waveform.

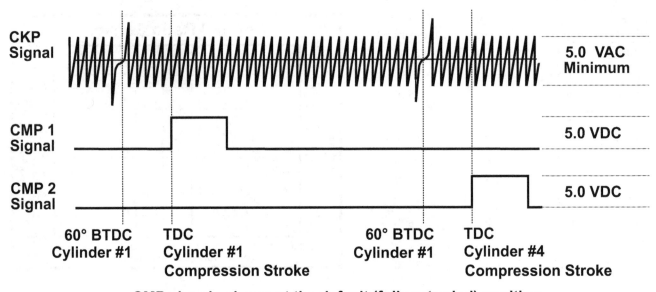

CMP signals shown at the default (fully retarded) position

EGR Valve Position Sensor

- A three-wire non-adjustable potentiometer that senses the position of the EGR valve pintle.
- Sensor output varies from 0.50 volts (valve fully closed) to 4.50 volts (valve fully open).
- Located on top of the EGR valve.

EGR Valve (% open)	Sensor Voltage
0	0.50
25	1.50
50	2.50
75	3.50
100	4.50

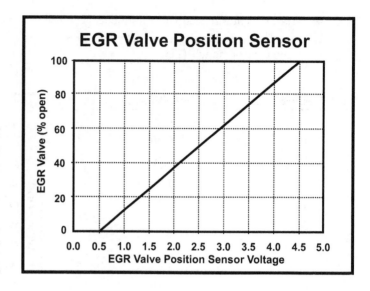

Fuel Level Sensor

- ■ A potentiometer that is used to determine the fuel level.
- ■ Sensor output varies from 0.5 volts at 0 % (empty tank) to 4.5 volts at 100 % (full tank).
- ■ Fuel tank at 1/4 full = 1.5 volts.
- ■ Fuel tank at 3/4 full = 3.5 volts.
- ■ Used by the ECM when testing the evaporative emission (EVAP) system.
- ■ Located in the fuel tank.

Fuel Level (% full)	Sensor Voltage
0	0.50
25	1.50
50	2.50
75	3.50
100	4.50

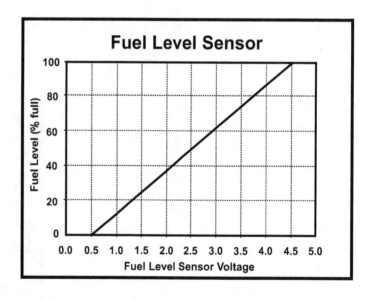

Fuel Pressure (FP) Sensor

- ■ A five-wire, combined, solid-state sensor used to monitor system pressure.
- ■ Located on the fuel rail. (Also see: Temperature Sensors > Fuel Temperature (FT) Sensor)
- ■ Sensor output varies from 0.50 volts at 0 psi to 4.50 volts at 90 psi. At 60 psi, the pressure sensor reading is 3.2 volts.

■ Measurement is referenced to atmosphere and will match mechanical gauge pressure.
■ Used by the ECM to measure fuel system pressure and as an input to determine command signal output to the fuel pump control module (FPCM).

Fuel Pressure (psi)	Sensor Voltage
0	0.50
10	0.80
20	1.40
30	1.75
40	2.25
50	2.75
60	3.20
70	3.65
80	4.10
90	4.50

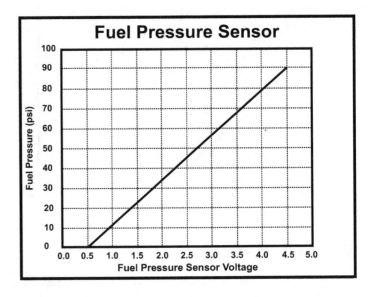

Fuel Tank (EVAP) Pressure Sensor

■ Senses vapor pressure or vacuum in the EVAP system compared to atmospheric pressure.
■ Sensor output varies from 0.5 volts at -0.500 psi (-14 in. H2O) when under vacuum, to 4.5 volts at 0.500 psi (14 in. H2O) when pressurized. With no pressure or vacuum in the fuel tank (fuel cap removed), the sensor output is 2.5 volts.
■ Used by the ECM for OBD evaporative emission system diagnostics only.
■ Located on top of the fuel tank.

Fuel Tank (EVAP) (in.H2O)	Pressure (psi)	Sensor Voltage
−14.0	−0.500	0.50
−10.5	−0.375	1.00
−7.0	−0.250	1.50
−3.5	−0.125	2.00
0.0	0.000	2.50
3.5	0.125	3.00
7.0	0.250	3.50
10.5	0.375	4.00
14.0	0.500	4.50

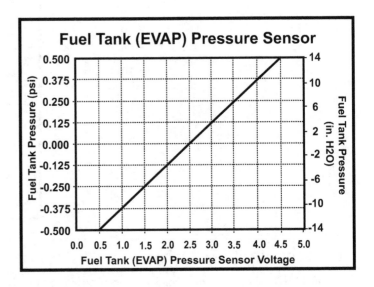

Knock Sensors

- Two-wire piezoelectric sensors that generate an AC voltage spike when engine vibrations within a specified frequency range are present.
- Located on each bank of the engine block.
- The signal is used by the ECM to retard ignition timing when knock is detected.
- The sensor signal circuit normally measures 2.5 volts DC with the sensor connected.

Heated Oxygen Sensor (HO2S 1/2 and HO2S 2/2)

- Electrically heated zirconia sensors.
- Mounted in the exhaust pipe (downstream) after the front catalytic converters on each bank.
- Used for OBD monitoring of catalytic converter efficiency.
- Sensor output varies from 0.0 to 1.0 volt.
- No bias voltage is applied to the sensor signal circuits by the ECM.
- With the key ON and engine OFF, the sensor reading is zero volts.
- Battery voltage is continuously supplied to the oxygen sensor heaters when the ignition switch is ON.
- Once the engine is started, the ECM will provide the ground for the downstream oxygen sensor heaters after two minutes of continuous engine operation.
- Normal oxygen sensor heater resistance is 8-12 Q at 68° F (20° C).

Ignition Switch

- Provides ignition key position input to the ECM.
- With the key in the RUN position and engine speed greater than 400 rpm, if an ignition switch fault is detected the engine will continue to run.

	IGN SW Pin a / ECM Pin 221	IGN SW Pin b / ECM Pin 222	IGN SW Pin c / ECM Pin 223	IGN SW Pin d / ECM Pin 224
OFF	0.0 V	B+	B+	B+
ACC	B+	0.0 V	B+	B+
RUN	B+	B+	0.0 V	B+
START	B+	B+	B+	0.0 V

Manifold Absolute Pressure (MAP) Sensor

- Senses intake manifold absolute pressure.
- Located on the intake manifold.
- Used by the ECM for OBD diagnostics and barometric pressure (BARO) calculation.
- ECM determines atmospheric altitude (BARO) during key ON/engine OFF.
- The normal BARO for the vehicle is sea level; 30 in. Hg (101 kPa).
- MAP sensor output varies between 4.5 volts (0 in. Hg vacuum / 101 kPa pressure) to 0.5 volts (24 in. Hg vacuum / 20.1 kPa pressure).
- Sensor output is 4.50 volts (0 in. Hg vacuum / 101 kPa pressure) at key ON/engine OFF at sea level.
- Sensor output is 1.17 volts at sea level with no load idle at 20 in. Hg vacuum (33.5 kPa pressure).
- ECM uses MAP input at wide open throttle (WOT) engine operation to update BARO measurement.

Vacuum at sea level (in. Hg. Gauge)	Manifold Absolute Pressure (kPa)	Sensor Voltage
0	101.3	4.50
3	91.2	4.00
6	81.0	3.50
9	70.8	3.00
12	60.7	2.50
15	50.5	2.00
18	40.4	1.50
21	30.2	1.00
24	20.1	0.50

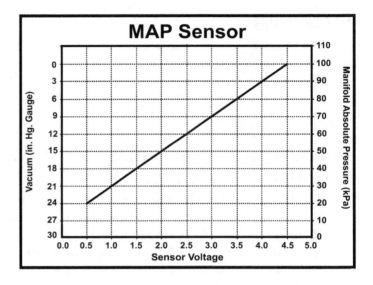

Mass Airflow (MAF) Sensor

- A hot-wire design that senses airflow into the intake manifold.
- Located on the air cleaner housing.
- Sensor output varies from 0.2 volts (0 gm/sec) at key ON/engine OFF, to 4.8 volts (175 gm/sec) at maximum airflow.
- At sea level, no-load idle (850 rpm), the sensor reading is 0.85 volts (3.0 gm/sec).

Mass Airflow (gm/sec)	Sensor Voltage
0	0.20
0	0.20
2	0.70
4	1.00
8	1.50
15	2.00
30	2.50
50	3.00
80	3.50
110	4.00
150	4.50
175	4.80

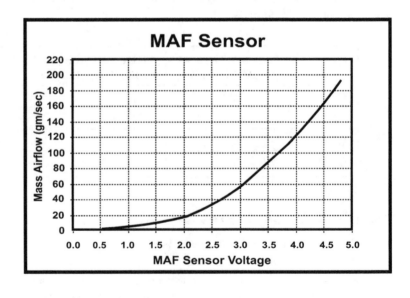

Throttle Position (TP 1 and TP 2) Sensors

- A pair of redundant non-adjustable potentiometers that sense throttle position.
- Located on the throttle body assembly.
- The TP 1 sensor output varies from 4.5 volts at closed throttle to 0.5 volts at maximum throttle opening (decreasing voltage with increasing throttle position).

- The TP 2 sensor signal varies from 0.5 volts at closed throttle to 4.5 volts at maximum throttle opening (increasing voltage with increasing throttle position).
- A circuit failure of one TP sensor will set a DTC and the ECM will limit the maximum throttle opening to 35 %.
- Circuit failure of both TP sensors, or a correlation error, will set a DTC, and will disable TAC.
- When disabled, the spring-loaded throttle plate returns to the default 15 % position (fast idle).

Throttle Position (% open)	TP 1 Sensor Voltage	TP 2 Sensor Voltage
0	4.50	0.50
5	4.30	0.70
10	4.10	0.90
15	3.90	1.10
20	3.70	1.30
25	3.50	1.50
40	2.90	2.10
50	2.50	2.50
60	2.10	2.90
75	1.50	3.50
80	1.30	3.70
100	0.50	4.50

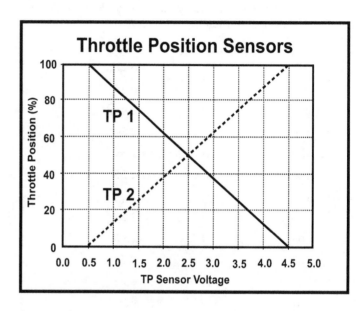

Temperature Sensors

Engine Coolant Temperature (ECT) Sensor

- A negative temperature coefficient (NTC) thermistor that senses engine coolant temperature.
- Located in the engine block water jacket.
- Measures temperatures from −40° F to 248° F (−40° C to 120° C).
- At 212° F (100° C), the sensor reading is 0.46 volts.

Fuel Temperature (FT) Sensor

- A five-wire, combined, solid-state sensor used to monitor fuel system temperature. A negative temperature coefficient (NTC) thermistor that senses fuel rail temperature.
- Located on the fuel rail. (Also see Fuel Pressure (FP) Sensor).
- Measures temperatures from −40° F to 248° F (−40° C to 120° C).
- At 86° F (30° C), the temperature sensor reading is 2.6 volts.
- The signal is used by the ECM to measure fuel system temperature and as an input to determine command signal output to the fuel pump control module (FPCM).

Intake Air Temperature (IAT) Sensor

- A negative temperature coefficient (NTC) thermistor that senses air temperature.
- Located in the air cleaner housing.
- Measures temperatures from −40° F to 248° F (−40° C to 120° C).
- At 86° F (30° C), the sensor reading is 2.6 volts.

Transmission Fluid Temperature (TFT) Sensor

- A negative temperature coefficient (NTC) thermistor that senses transmission fluid temperature.
- Located in the transaxle oil pan.
- Measures temperatures from −40° F to 248° F (−40° C to 120° C).
- At 212° F (100° C), the sensor reading is 0.46 volts.
- This signal is used by the TCM to delay shifting when the fluid is cold, and control torque converter clutch operation when the fluid is hot.

Temperature °F	Temperature °C	Sensor Voltage
248	120	0.25
212	100	0.46
176	80	0.84
150	66	1.34
140	60	1.55
104	40	2.27
86	30	2.60
68	20	2.93
32	0	3.59
−4	−20	4.24
−40	−40	4.90

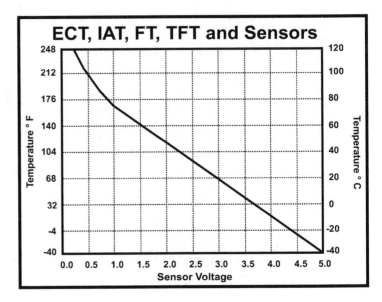

Transmission Range (TR) Switch

- A five-position switch that indicates the position of the transaxle manual select lever: PARK, REVERSE, NEUTRAL, DRIVE (D), or LOW (L).
- Located on the transaxle housing.
- Used by the TCM to control line pressure, upshifting, and downshifting.

Transmission Turbine Shaft Speed (TSS) Sensor

- A magnetic-type sensor that senses rotation of the torque converter turbine shaft (input/mainshaft).
- Located on the transaxle housing.
- Generates a signal that increases in frequency as transmission input speed increases.
- Used by the TCM to control torque converter clutch operation and sense transmission slippage.

Variable Valve Lift System Sensors

- Two, three-wire, non-adjustable potentiometers that sense the variable valve lift motor shaft position for each variable valve lift motor.
- Located in the variable valve lift motor assemblies.
- Sensor output will read 0.50 V at low lift (0 % command) and 4.50 V at high lift (100 % command).
- Normal ECM command of the motors at idle is base lift. At wide open throttle (WOT), maximum load, the ECM commands maximum additional lift.

Vehicle Speed Sensor (VSS)

- A magnetic-type sensor that senses rotation of the final drive.
- Located on the transaxle housing.
- Generates a signal that increases in frequency as vehicle speed increases.
- The TCM uses the VSS signal to control upshifts, downshifts, and the torque converter clutch.
- The TCM communicates the VSS signal over the data communications bus to the ECM to control high speed fuel cutoff, and to the Instrument Cluster for speedometer operation.
- The signal is displayed on the scan tool in miles per hour and kilometers per hour.

OUTPUTS - ACTUATORS

Camshaft Position Actuator Control Solenoids

- A pair of duty-cycle controlled solenoid valves that modify the valve timing of the intake camshafts by controlling engine oil flow to the camshaft position actuators.
- As duty cycle increases, oil flows from the solenoid to the actuator advancing the camshaft position.
- As the duty cycle decreases, the amount of oil flow from the solenoid is reduced allowing the camshaft to move back towards the rest position.
- When the ECM determines that the desired camshaft position has been achieved, the duty cycle is commanded to 50 % to hold the actuator so that the adjusted camshaft position is maintained.
- The solenoid winding resistance specification is $12 \pm 2\ \Omega$.

Evaporative Emission (EVAP) Canister Purge Solenoid

- Duty cycle controlled regulation of EVAP canister purge vapor flow into the intake manifold.
- Enabled when the engine coolant temperature reaches 150° F (66° C).
- A duty cycle of 0 % blocks vapor flow, and a duty cycle of 100 % allows maximum vapor flow.
- The duty cycle is determined by the ECM, based on engine speed and load.
- Also used for OBD testing of the evaporative emission (EVAP) system.
- A service port with a schrader valve is on the hose between the purge solenoid and the canister.
- Winding resistance specification is $36 \pm 4\ \Omega$.

Evaporative Emission (EVAP) Canister Vent Solenoid

- When energized, the fresh air supply hose to the canister is blocked.
- Energized only during OBD testing of the evaporative emission (EVAP) system.
- Winding resistance specification is $36 \pm 4\ \Omega$.

Exhaust Gas Recirculation (EGR) Valve

- A duty cycle controlled solenoid that controls the spring-loaded EGR valve pintle.
- A scan tool value of 0 % indicates an ECM command to fully close the EGR valve.
- A scan tool value of 100 % indicates an ECM command to fully open the EGR valve.
- Enabled when the engine coolant temperature reaches 150° F (66° C) and the throttle is not closed or wide open.
- Winding resistance specification is $12 \pm 2\ \Omega$.

Fan Control (FC) Relay

- When energized, the relay provides battery voltage (B+) to the radiator/condenser cooling fan motor.
- Energized when engine coolant temperature reaches 220° F (104° C); off when coolant temperature drops to 195° F (90° C).

- Energized when the A/C high side pressure reaches 300 psi (2068 kPa); off when the pressure drops to 250 psi (1724 kPa).
- Coil resistance specification is $36 \pm 4 \, \Omega$.

Fuel Injectors

- Electromechanical devices used to deliver fuel to the intake manifold at each cylinder.
- Each individually energized once per camshaft revolution, in time with its cylinder's exhaust stroke.
- Winding resistance specification is $12 \pm 2 \, \Omega$.

Generator

- The ECM supplies a variable duty-cycle signal to ground the field winding of the generator (alternator).
- The ECM receives battery/charging voltage input at pin 219. This pin is a dedicated generator input.
- Increased duty cycle results in a higher field current and greater generator (alternator) output.

Ignition Coils

- Coil-on-plug (COP) system with six individual coils connected directly to the spark plugs.
- Timing and dwell are controlled by the ECM.
- Coil primary resistance specification is $1 \pm 0.5 \, \Omega$.
- Coil secondary resistance specification is $10K \pm 2K \, \Omega$.

Malfunction Indicator Lamp (MIL)

- Part of the instrument cluster module (ICM).
- Receives commands from the ECM and TCM over the data communications bus.
- If the ICM is unable to communicate with the communications bus network, the MIL will be lit.
- With no faults present, the MIL is lit for 5 seconds after the ignition switch is turned ON (bulb check).
- An emissions-related fault is present if the MIL stays lit after the bulb check.
- When misfiring occurs that could damage a catalytic converter, the MIL flashes on and off.

Starter Relay

- When energized, provides battery voltage (B+) to the starter solenoid.
- Energized based upon ignition switch position (START), transmission range switch position (PARK/ NEUTRAL), vehicle speed (0 mph), and engine speed (0 rpm).
- Coil resistance specification is $36 \pm 4 \, \Omega$.

Throttle Actuator Control (TAC) Motor

- A bidirectional pulse-width modulated DC motor that controls the position of the throttle plate.
- Scan tool data value of 0 % = ECM command to fully close throttle plate.
- Scan tool value of 100 % = ECM command to fully open the throttle plate (wide open throttle).
- Any throttle control actuator motor circuit fault sets a DTC and causing the TAC to be disabled, and the spring-loaded throttle plate will return to the default 15 % position (fast idle).
- When disabled, the TAC value on the scan tool will indicate 15 %.
- Maximum throttle actuator control motor current is 6 amps.

Transmission Solenoids

Torque Converter Clutch (TCC) Solenoid Valve

- This normally low (NL) variable force solenoid controls fluid in the transmission valve body that is routed to the torque converter clutch.
- TCM varies duty cycle to maintain a controlled slip or a full application of the clutch (zero slip).
- Scan tool duty cycle value of 0 % = TCC is released.
- When torque converter clutch application is desired, the pulse width increases.
- Scan tool duty cycle value of 100 % = TCC is fully applied.
- The duty cycle is immediately cut to 0 % (released) if the brake pedal position switch closes.
- Enabled when the engine coolant temperature reaches 150° F (66° C), the brake switch is open, the transmission is in 3rd gear or higher, and the vehicle is at cruise (steady throttle) above 35 mph.
- Winding resistance specification is 6 ± 1 Ω.

Transmission Electronic Pressure Control (EPC) Solenoid

- This normally high (NH) variable force solenoid controls fluid in the transmission valve body that is routed to the pressure regulator valve.
- TCM varies duty cycle to modify the line pressure of the transmission for best shift quality.
- Scan tool duty cycle value of 10 % = maximum line pressure command.
- Scan tool duty cycle value of 90 % = minimum line pressure is commanded.
- Winding resistance specification is 6 ± 1 Ω.

Transmission Shift Solenoids (SS A, SS B, SS C, SS D, and SS E)

- Control fluid flow to the clutches.
- Located in the transmission valve body.
- SS A and SS D are normally low (NL) variable force solenoids.
- SS B and SS C are normally high (NH) variable force solenoids.
- SS E is an OFF/ON solenoid that is normally closed (NC).
- By modifying the duty cycle of the variable force solenoids and changing the state of the ON/OFF solenoid, the TCM can control the pressure to the clutches to enable a gear change.
- Winding resistance specification is 6 ± 1 Q, except for SS E which is 22 ± 2 Ω.

Gear Selector Position	PCM Gear Command	4-5-6 Clutch	3-5-R Clutch	2-6 Clutch	L-R Clutch	1-2-3-4 Clutch	Low/ One-Way Clutch	SS A (VFS) 1-2-3-4 Clutch NL	SS B (VFS) L-R/ 4-5-6 NH	SS C (VFS) 3-5-R NH	SS D (VFS) 2-6 NL	SS E (OFF/ON) NC	Gear Ratio
P	P				Applied			OFF	OFF	ON	OFF	ON	
R	R		Applied		Applied			OFF	OFF	OFF	OFF	ON	2.88:1
N	N				Applied			OFF	OFF	ON	OFF	ON	
D	1				Applied	Applied	Applied	ON	OFF	ON	OFF	ON	4.48:1
	2			Applied		Applied	OR	ON	ON	ON	ON	OFF	2.87:1
	3		Applied			Applied	OR	ON	ON	OFF	OFF	OFF	1.84:1
	4	Applied				Applied	OR	ON	OFF	ON	OFF	OFF	1.41:1
	5	Applied	Applied				OR	OFF	OFF	OFF	OFF	OFF	1:1
	6	Applied		Applied			OR	OFF	OFF	ON	ON	OFF	0.74:1
L	L				Applied	Applied		ON	OFF	ON	OFF	ON	4.48:1

VFS = Variable Force Solenoid   NL = Normally Low   OR = Over Running
NC = Normally Closed   NH = Normally High

Normally Low (NL): When this solenoid type is O FF (not energized), fluid pressure is low, fluid is exhausted from the circuit, and the clutch is not applied. When the solenoid is ON (energized), the TCM increases the fluid pressure by varying the solenoid duty cycle and the clutch is applied.

Normally High (NH): When this solenoid type is OFF (not energized), fluid pressure is high in the fluid circuit and the clutch is applied. When the solenoid is ON (energized), the TCM reduces the fluid pressure by varying the solenoid duty cycle and the clutch is released.

Normally Closed (NC): This solenoid type is either ON or OFF and controls a three-port hydraulic circuit to aid in shift strategy. Regulated line pressure is switched between two hydraulic ports, a default passage, and a primary passage. When this solenoid type is OFF (not energized), oil is blocked from the primary oil passage and fed to a default passage. When the solenoid is ON (energized), the TCM redirects the fluid pressure to the primary oil passage.

Variable Valve Lift Motors

- A bidirectional DC motor that controls the position of the variable valve lift mechanism.
- DC motor is attached to a rod which operates a fulcrum attached to the rocker arms.
- Changes in rocker arm ratio result in additional lift above the base lift provided by the camshaft lobes.
- Scan tool value below 3000 rpm = 0.5 V (0 %).
- Scan tool value above 3000 rpm = 4.5 V (100 %).

OBD SYSTEM OPERATION

Data Communications

Powertrain Communications Network

- High-speed, serial data bus.
- Two-wire twisted pair communications network.
- Allows peer-to-peer communications between the ECM, TCM, instrument cluster (including the MIL), immobilizer control module, and a scan tool connected to the data link connector (DLC).
- Data-High circuit switches between 2.5 (rest state) and 3.5 volts (active state).
- Data-Low circuit switches between 2.5 (rest state) and 1.5 volts (active state).
- Two, 120-ohm terminating resistors: one inside the instrument cluster and another inside the ECM.
- Any of the following conditions will cause serial data bus communications to fail and result in the storage of network DTCs:
  - either data line shorted to voltage.
  - either data line shorted to ground.
  - one data line shorted to the other data line.
  - an open in either data line to a module.
- Data bus remains operational when one of the two modules containing a terminating resistor is not connected to the network.
- Data bus will fail when both terminating resistors are not connected to the network.
- Communication failures will not prevent the ECM from providing control of the ignition system.

Fuel Pump Control Module (FPCM) Communications Network

- Local Area Network (LAN) bus.
- Two-wire, twisted pair communications network; isolated from the powertrain communications network.
- Allows peer-to-peer communications between the ECM and the FPCM only.
- LAN Data-High circuit switches between 2.5 (rest state) and 3.5 volts (active state).
- LAN Data-Low circuit switches between 2.5 (rest state) and 1.5 volts (active state).

- Any of the following conditions will cause LAN data bus communications to fail:
  - either data line shorted to voltage.
  - either data line shorted to ground.
  - either data line open.
  - one data line shorted to the other data line.

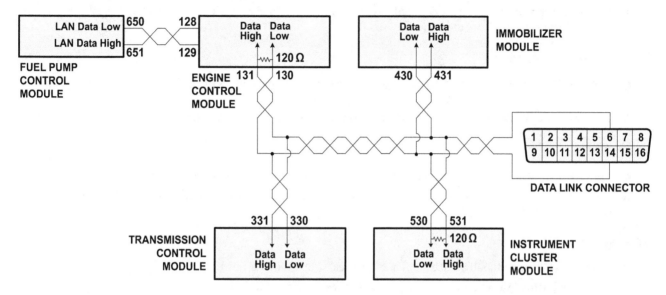

System Monitors

The OBD diagnostic system also actively tests some systems for proper operation while the vehicle is being driven. Fuel control and engine misfire are checked continuously. Air/fuel ratio sensor response, air/fuel ratio sensor heater operation, oxygen sensor response, oxygen sensor heater operation, catalyst efficiency, EGR operation and EVAP integrity are tested once or more per trip. When any of the System Monitors detects a failure that will result in emissions exceeding a predetermined level on two consecutive trips, the ECM will store a diagnostic trouble code (DTC) and illuminate the malfunction indicator lamp (MIL). Freeze frame data captured during the first of the two consecutive failures is also stored.

Air/Fuel Ratio & Oxygen Sensors

- Checks the maximum and minimum signal output and response times for all air/fuel ratio sensors and oxygen sensors.
- If an air/fuel ratio sensor or oxygen sensor signal remains too low, too high, responds too slowly, or does not respond, a DTC is set.

Air/Fuel Ratio & Oxygen Sensor Heaters

- Checks the current flow through each air/fuel ratio sensor heater and the oxygen sensor heater.
- If the current flow is too high or too low, a DTC is set.
- Battery voltage is continuously supplied to the air/fuel ratio sensor heaters and oxygen sensor heaters whenever the ignition switch is on.
- The heaters are grounded through the ECM.

Catalytic Converter

- Compares the data from the heated air/fuel ratio sensors (upstream) to the heated oxygen sensors (downstream) to determine the oxygen storage capability of the catalysts.
- If a catalyst's oxygen storage capacity is sufficiently degraded, the ECM will store the appropriate DTC and illuminate the MIL.

■ Will run only after the air/fuel ratio sensor heater, oxygen sensor heater, air fuel ratio sensor and oxygen sensor monitors have run and passed.

Comprehensive Component

■ Continuous monitor of all engine and transmission sensors and actuators for shorts and opens, as well as values that do not logically fit with other powertrain data (rationality).

■ On the first trip where a failure is detected, the ECM or TCM will store a DTC. The ECM will then store a freeze frame of data and illuminate the MIL.

EGR System

■ Uses the MAP sensor signal to detect changes in intake manifold pressure as the EGR valve is commanded open and closed.

■ If the pressure changes too little or too much as compared to the EGR valve position sensor input, a DTC is set.

Engine Misfire

■ Uses the CKP sensor signal to continuously detect engine misfires, both severe and non-severe.

■ If the misfire is severe enough to cause catalytic converter damage, the MIL will flash on and off as long as the severe misfire is detected.

EVAP System

■ Tests for small leaks (0.020 in./0.5 mm) and large leaks (.040 in./1.0 mm).

■ Engine off, natural vacuum leak detection is used to test for a small leak (0.020 in./0.5 mm).

    ■ Enable criteria for the small leak test:

        ■ the vehicle must have been driven between 15 to 90 minutes.

        ■ fuel level must be between 1/4 and 3/4 full.

        ■ ambient air temperature must be between 40° F (4.4° C) and 105° F (40° C).

        ■ the key is OFF/the engine is OFF.

    ■ When the key is turned OFF, the vent solenoid is left open for ten minutes to allow the system to stabilize. The ECM then notes the fuel tank pressure (FTP).

    ■ The ECM then energizes the EVAP vent solenoid for four minutes while monitoring the fuel tank pressure (FTP) sensor for a pressure change.

    ■ If the system reaches the target value, a change of greater than 1.0 in. H2O from the stabilized reading, the test is complete and the system passes (no leak detected).

    ■ If the system fails to pass the initial small leak test, the ECM will then command the EVAP vent solenoid open for two minutes. The ECM then notes the fuel tank pressure (FTP).

    ■ The ECM then energizes the EVAP vent solenoid for 20 minutes while monitoring the FTP sensor for change.

    ■ A change in fuel tank pressure of greater than 1.0 in. H2O indicates a pass (no leak detected).

■ Vacuum decay is used to test for a large leak (.040 in./1.0 mm).

    ■ Enable criteria for the large leak test:

        ■ a cold start with engine temperature below 86° F (30° C).

        ■ fuel level must be between 1/4 and 3/4 full.

        ■ ambient air temperature must be between 40° F (4.4° C) and 105° F (40° C).

        ■ the engine is running.

    ■ The ECM turns on the EVAP vent solenoid, blocking the fresh air supply to the EVAP canister.

    ■ The EVAP purge solenoid is turned on to draw a slight vacuum on the entire EVAP system, including the fuel tank.

    ■ Then the EVAP purge solenoid is turned off to seal the system.

    ■ The monitor uses the Fuel Tank (EVAP) Pressure Sensor signal to determine if the EVAP system has any leaks.

    ■ After the testing is completed, the EVAP vent solenoid is turned off to relieve the vacuum.

- A small leak DTC will set if a sufficient change in fuel tank pressure is not achieved during the small leak test.
- A large leak DTC will set if sufficient vacuum is not created, or decays too rapidly, or does not decay quickly at the conclusion of the large leak test.

FUEL CONTROL

- Uses fuel trim and loop status to determine failures in the fuel system.
- Sets a DTC if the system fails to enter Closed Loop mode within 2 minutes of startup.
- Sets a DTC if Long Term Fuel Trim reaches its limit (+30 % or −30 %) indicating a loss of fuel control.

Monitor Readiness Status

- Indicates whether or not the OBD diagnostic monitor has completed.
- If the monitor has not completed, the status on the scan tool displays "NOT COMPLETE."
- If the monitor has completed, the status on the scan tool displays "COMPLETE."
- When DTCs are cleared from memory or the battery is disconnected, all non-continuous monitors will have the readiness status indicators reset to "NOT COMPLETE."
- The readiness status of the following non-continuous system monitors can be read on the scan tool:
  - Catalytic Converter
  - EGR System
  - EVAP System
  - Oxygen Sensors
  - Oxygen Sensor Heaters

Warm Up Cycle

- Used by the ECM and TCM for automatic clearing of DTCs and Freeze Frame data (described below).
- Must have an increase of at least 40° F (an increase of 22° C) and reach a minimum of 160° F (71° C).

Trip

- A key-on cycle in which all enable criteria for a diagnostic monitor are met and the monitor is run.
- The trip completes when the ignition switch is turned off.

Drive Cycle

Most OBD monitors will run at some time during normal operation of the vehicle. However, to satisfy all of the different Trip enable criteria and run all of the OBD diagnostic monitors, the vehicle must be driven under a variety of conditions. The following drive cycle will allow all monitors to run on this vehicle.

- Ensure that the fuel tank is between 1/4 and 3/4 full.
- Engine cold start below 86° F (30° C).
- Engine warm up until coolant temperature is at least 160° F (71° C).
- Accelerate to 40-55 mph at 25 % throttle and maintain speed for 5 minutes.
- Decelerate without using the brake (coast down) to 20 mph or less, and then stop the vehicle. Allow the engine to idle for 10 seconds, turn the key off, and wait 1 minute.
- Restart and accelerate to 40-55 mph at 25 % throttle and maintain speed for 2 minutes.
- Decelerate without using the brake (coast down) to 20 mph or less, and then stop the vehicle. Allow the engine to idle for 10 seconds, turn the key off, and wait 45 minutes.

Freeze Frame Data

- A snapshot (one frame of data) that is automatically stored in the memory of either the ECM or TCM when an emission-related DTC is first stored (pending).

- If a DTC for fuel control or engine misfire is stored at a later time, the newest data is stored, replacing the earlier data.
- All parameter ID (PID) values listed under "Scan Tool Data" are stored in freeze frame data.
- The ECM and TCM store only one single freeze frame record.

Storing and Clearing DTCS & Freeze Frame Data, Turning the Mil On & Off

One Trip Monitors

- A failure on the first trip of a "one trip" emissions diagnostic monitor causes the ECM or TCM to immediately store a confirmed DTC, capture Freeze Frame data, and turn on the MIL.
- All Comprehensive Component Monitor faults set a confirmed DTC on one trip.

Two Trip Monitors

- A failure on the first trip of a "two trip" emissions diagnostic monitor causes the ECM to store a pending DTC and Freeze Frame data.
- Normally, if the failure recurs on the next trip during which the monitor runs, regardless of the engine conditions, the ECM will store a confirmed DTC and turn on the MIL.
- For the misfire and fuel control monitors, if the failure recurs on the next trip during which the monitor runs and where conditions are similar to those experienced when the fault first occurred (engine speed within 375 rpm, engine load within 20 %, and same hot/cold warm-up status), the ECM will store a confirmed DTC and turn on the MIL.
- If the second failure does not recur as described above, the pending DTC and Freeze Frame data are cleared from memory.
- All of the System Monitors are two trip monitors.
- Engine misfire which is severe enough to damage the catalytic converter is a two trip monitor. The MIL will always flash on and off when the severe misfire is occurring.

Automatic Clearing

- When the vehicle completes three consecutive good/passing trips (three consecutive trips in which the monitor that set the DTC is run and passes), the MIL will be turned off, but the confirmed DTC and Freeze Frame will remain stored in ECM/TCM memory.
- For misfire and fuel control monitor faults, the three consecutive good/passing trips must take place under similar engine conditions (engine speed, load, and warm up condition) as the initial fault for the MIL to be turned off.
- If the vehicle completes 40 Warm Up cycles without the same fault recurring, the DTC and Freeze Frame are automatically cleared from the ECM/TCM memory.

Manual Clearing

- Any stored DTCs and Freeze Frame data can be erased using the scan tool, and the MIL (if lit) will be turned off.
- Although not the recommended method, DTCs and Freeze Frame data will also be cleared if the B+ power supply for the ECM/TCM is lost, or the battery is disconnected.

Scan Tool

- Can be used to communicate with the ECM, TCM, Immobilizer, and Instrument Cluster modules.
- Module reprogramming and initialization can be performed using the scan tool.
- The ECM, TCM, and instrument cluster are equipped with software that allows requests to be made through the OBD scan tool for output control of components and functional testing of systems.

Note: All testing is performed at sea level unless otherwise indicated.

PIN / Component Cross Reference

PIN#	Abbreviation	Diagram	PIN#	Abbreviation	Diagram	PIN#	Abbreviation	Diagram
1	+5 V	1 of 4	155	AFRS 2/1 Heater	2 of 4	340	TSS +	4 of 4
2	Ign	1 of 4	160	2S 1/2+	2 of 4	341	TSS -	4 of 4
3	B+	1 of 4	161	HO2S 1/2-	2 of 4	343	P	4 of 4
9	A/C Clutch	1 of 4	162	HO2S 1/2 Heater	2 of 4	344	R	4 of 4
10	Fan Control	1 of 4	163	HO2S 2/2+	2 of 4	345	N	4 of 4
14	Coil 1	1 of 4	164	HO2S 2/2-	2 of 4	346	D	4 of 4
15	Coil 3	1 of 4	165	HO2S 2/2 Heater	2 of 4	347	L	4 of 4
16	Coil 5	1 of 4	171	BPP	2 of 4	348	+5 V	4 of 4
17	Coil 2	1 of 4	172	A/C Request	2 of 4	350	Sensor Ground	4 of 4
18	Coil 4	1 of 4	181	MAF	2 of 4	360	Ground	4 of 4
19	Coil 6	1 of 4	201	Starter Control	3 of 4	402	Connector #12	4 of 4
22	TAC	1 of 4	203	CMP 1 Sol	3 of 4	405	Ant. +	4 of 4
23	TAC	1 of 4	204	CMP 2 Sol	3 of 4	406	Ant. -	4 of 4
24	TP 1	1 of 4	205	VVLB1 A	3 of 4	430	Data Low	4 of 4
25	TP 2	1 of 4	206	VVLB1 B	3 of 4	431	Data High	4 of 4
28	EGR Position	1 of 4	207	̄ V	3 of 4	460	Ground	4 of 4
29	MAP	1 of 4	208	VVLS 1	3 of 4	502	Connector #25	4 of 4
30	FPS	1 of 4	209	Sensor Ground	3 of 4	530	Data Low	4 of 4
31	FTS	1 of 4	210	VVLB2 A	3 of 4	531	Data High	4 of 4
34	ECT	1 of 4	211	VVLB2 B	3 of 4	560	Ground	4 of 4
35	IAT	1 of 4	212	+5 V	3 of 4	605	B+	2 of 4
40	A/C Pressure	1 of 4	213	VVLS 2	3 of 4	610	FP Feed	2 of 4
41	Fuel Tank Pressure	1 of 4	214	Sensor Ground	3 of 4	611	FP Ground	2 of 4
42	Fuel Level	1 of 4	219	Battery Sense	3 of 4	649	FP Enable	2 of 4
50	Sensor Ground	1 of 4	221	IGN OFF	3 of 4	650	LAN Data Low	2 of 4
60	Ground	1 of 4	222	IGN ACC	3 of 4	651	LAN Data High	2 of 4
101	Gen Field	2 of 4	223	IGN RUN	3 of 4	660	Ground	2 of 4
104	EVAP Vent	2 of 4	224	IGN CRANK	3 of 4			
108	EGR	2 of 4	240	KS B1	3 of 4	**Component**		**Diagram**
110	EVAP Purge	2 of 4	241	KS B2	3 of 4	Connector 1		1
120	INJ 1	2 of 4	250	CMP 1	3 of 4	Connector 8		4
121	INJ 1	2 of 4	251	CMP 2	3 of 4	Connector 12		4
122	INJ 3	2 of 4	252	CKP +	3 of 4	Connector 25		4
123	INJ 4	2 of 4	253	CKP -	3 of 4			
124	INJ 5	2 of 4	302	Ign.	4 of 4	**Component**		**Diagram**
125	INJ 6	2 of 4	303	B+	4 of 4	Fuse 3		1
130	Data Low	2 of 4	305	TCC	4 of 4	Fuse 4		1
131	Data High	2 of 4	306	EPC	4 of 4	Fuse 20		2
140	+5 V	2 of 4	307	SS A	4 of 4	Fuse 22		2
141	APP 1	2 of 4	308	SS B	4 of 4	Fuse 30		3
142	Sensor Ground	2 of 4	309	SS C	4 of 4	Fuse 31		3
143	+5 V	2 of 4	310	SS D	4 of 4	Fuse 32		3
144	APP 2	2 of 4	311	SS E	4 of 4	Fuse 34		3
145	Sensor Ground	2 of 4	330	Data Low	4 of 4	Fuse 36		3
150	AFRS 1/1+	2 of 4	331	Data High	4 of 4	Fuse 40		4
151	AFRS 1/1-	2 of 4	336	TFT	4 of 4	Fuse 41		4
152	AFRS 1/1 Heater	2 of 4	337	VSS +	4 of 4	Fuse 42		4
153	AFRS 2/1+	2 of 4	338	VSS -	4 of 4	Fuse 43		4
154	AFRS 2/1-	2 of 4				Fuse 44		4

The information on these two pages are the data values, including minimum-to-maximum ranges, that the OBD II scan tool is capable of displaying for each of the data parameters.

Absolute Load Value	0 to 100 %
A/C Clutch	ON / OFF
A/C Pressure	25 to 450 psi / 0.0 to 5.0 V
A/C Request	ON / OFF
AFRS 1/1	-1.00 to 5.00 V
AFRS 2/1	-1.00 to 5.00 V
AFRS 1/1 Current	-9999 to +9999 microamps
AFRS 2/1 Current	-9999 to +9999 microamps
AFRS 1/1 Heater	0 to 100 %
AFRS 2/1 Heater	0 to 100 %
Air/Fuel Lambda Bank 1	0.00 to 2.00
Air/Fuel Lambda Bank 2	0.00 to 2.00
APP 1	0 to 100 % / 0.0 to 5.0 V
APP 2	0 to 100 % / 0.0 to 5.0 V
BARO	101 to 67 kPa pressure / 30 to 20 in.Hg. pressure
Battery Voltage	0 to 18 V
Brake Pedal Position Switch	ON / OFF
Cam 1 Desired Advance	0 to 99°
Cam 2 Desired Advance	0 to 99°
Cam 1 Solenoid Control	0 to 100 %
Cam 2 Solenoid Control	0 to 100 %
CMP 1	0° to 99°
CMP 2	0° to 99°
Distance traveled since DTCs cleared	#### miles/km
Distance traveled with MIL on	#### miles/km
DTCs (confirmed)	P####, U####, etc.
DTCs (pending)	P####, U####, etc.
ECT	248 to -40° F / 120 to -40° C / 0.0 to 5.0 V
EGR Position Sensor	0 to 100 % / 0.0 to 5.0 V
EGR Valve Opening Desired	0 to 100 %
Electronic Pressure Control (EPC)	0 to 100 %
Engine RPM	0 to 9999 rpm
Evap Purge Solenoid	0 to 100 %
Evap Vent Solenoid	ON / OFF
Fan Control	ON / OFF
Fuel Enable	YES / NO
Fuel Pressure	0.00 to 5.00 V / 0 to 90 psi
Fuel Pump Command (ECM)	OFF / LOW / HIGH
Fuel Pump Feedback (FPCM)	0 to 100 %
Fuel Tank (EVAP) Pressure	-14.0 to +14.0 in.H2O / -0.5 psi to 0.5 psi / 0.0 to 5.0 V
Fuel Tank Level	0 to 100 % / 0.0 to 5.0 V
Fuel Temperature	248 to -40° F / 120 to -40° C / 0.0 to 5.0 V
Generator Field	0 to 100 %

The information on these two pages are the data values, including minimum-to-maximum ranges, that the OBD II scan tool is capable of displaying for each of the data parameters.

HO2S 1/2	-1.00 to 2.00 V
HO2S 2/2	-1.00 to 2.00 V
HO2S 1/2 Heater	ON / OFF
HO2S 2/2 Heater	ON / OFF
IAT	248 to -40° F / 120 to -40° C / 0.0 to 5.0 V
Ignition Switch	OFF / ACC / RUN / START
Ignition Timing Advance	-99° to 99° BTDC
Injector Pulse Width Bank 1	0 TO 99 ms
Injector Pulse Width Bank 2	0 to 99 ms
Knock Sensor B1 (knock detected)	YES / NO
Knock Sensor B2 (knock detected)	YES / NO
Long Term Fuel Trim Bank 1	-99 % to +99 %
Long Term Fuel Trim Bank 2	-99 % to +99 %
Loop Status	OPEN / CLOSED
MAF	0 to 175 gm/sec / 0.0 to 5.0 V
MAP	20 to 101 kPa pressure / 24 to 0 in.Hg. vacuum / 0.0 to 5.0 V
MIL	ON / OFF / FLASHING
Monitor Status for this trip	DISABLED / NOT COMPLETE / COMPLETE
Number warm-up cycles since DTCs cleared	###
Shift Solenoid A	0 to 100 %
Shift Solenoid B	0 to 100 %
Shift Solenoid C	0 to 100 %
Shift Solenoid D	0 to 100 %
Shift Solenoid E	ON / OFF
Short Term Fuel Trim Bank 1	-99 % to +99 %
Short Term Fuel Trim Bank 2	-99 % to +99 %
Software Calibration ID Number	(CAL ID)
Software Verification Number	(CVN)
Starter Relay	ON / OFF
TCC	0 to 100 %
TFT	248 to -40° F / 120 to -40° C / 0.0 to 5.0 V
Throttle Actuator Control	0 to 100 %
Time elapsed since engine start	hh:mm:ss
TP 1	0 to 100 % / 0.0 to 5.0 V
TP 2	0 to 100 % / 0.0 to 5.0 V
TR	P, N, R, D, L
TSS	0 to 9999 rpm
Valid Ignition Key	YES / NO
Variable Valve Lift Bank 1 Commanded	ON / OFF
Variable Valve Lift Bank 2 Commanded	ON / OFF
Variable Valve Lift Position Bank 1	0 to 100 % / 0.0 to 5.0 V
Variable Valve Lift Position Bank 2	0 to 100 % / 0.0 to 5.0 V
Vehicle Identification Number	(VIN)
VSS	0 to 199 mph

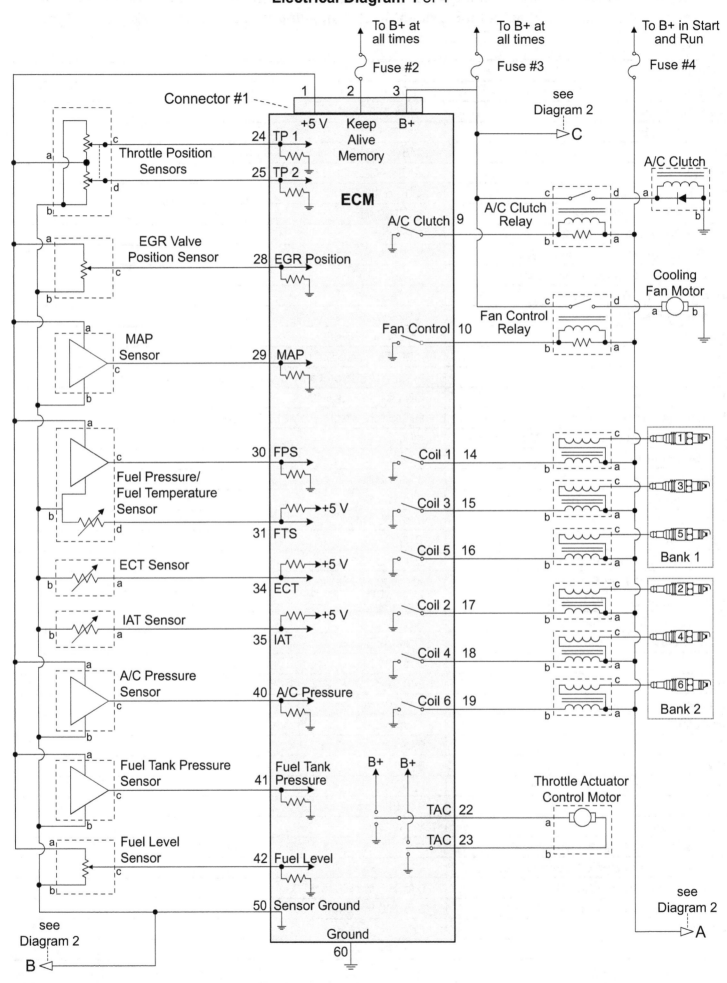

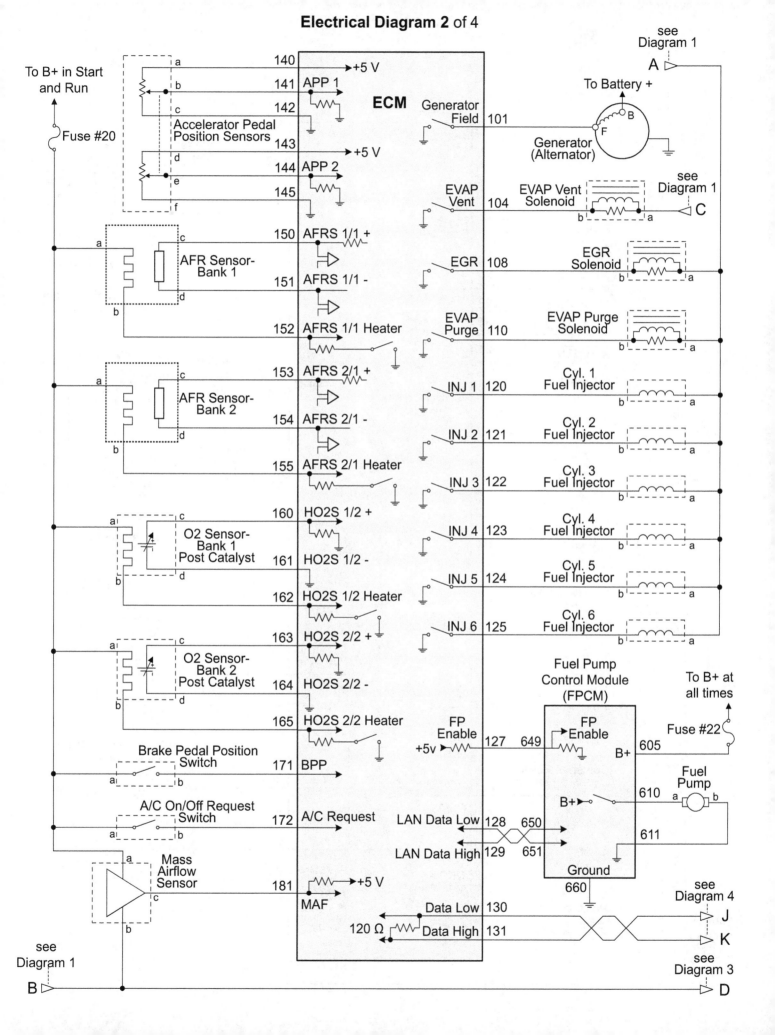

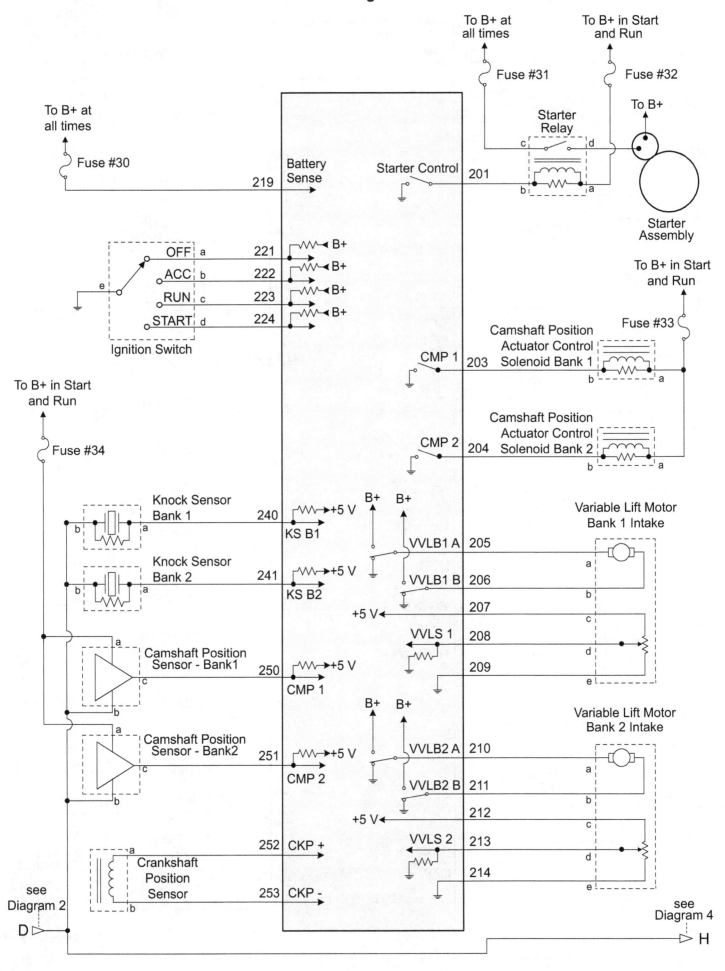

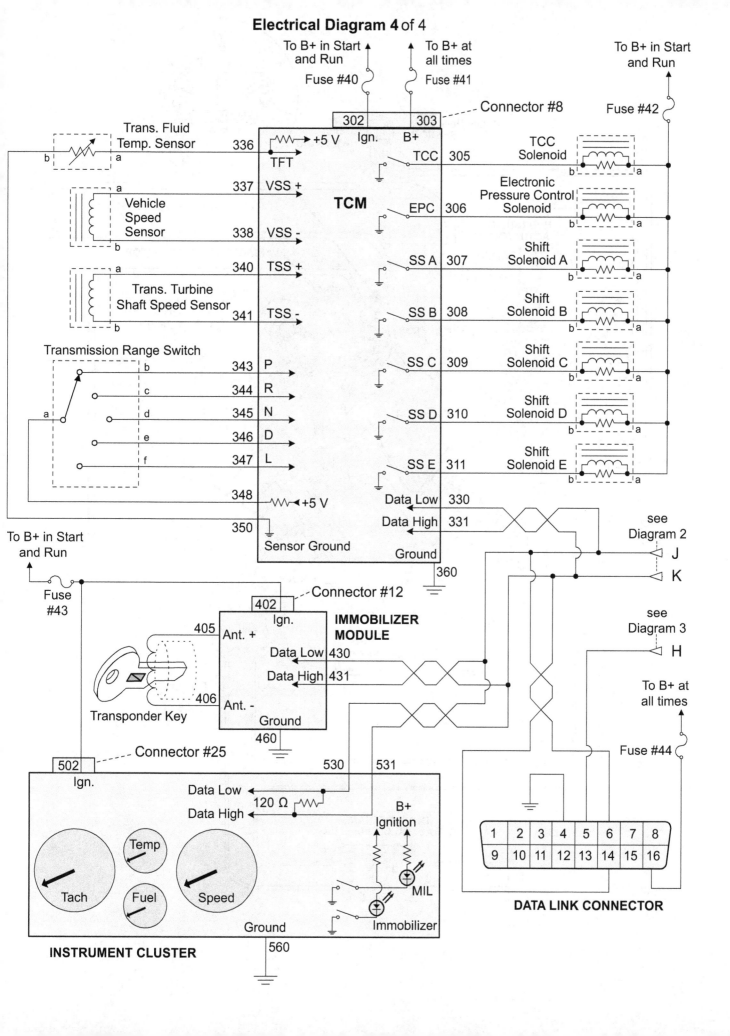

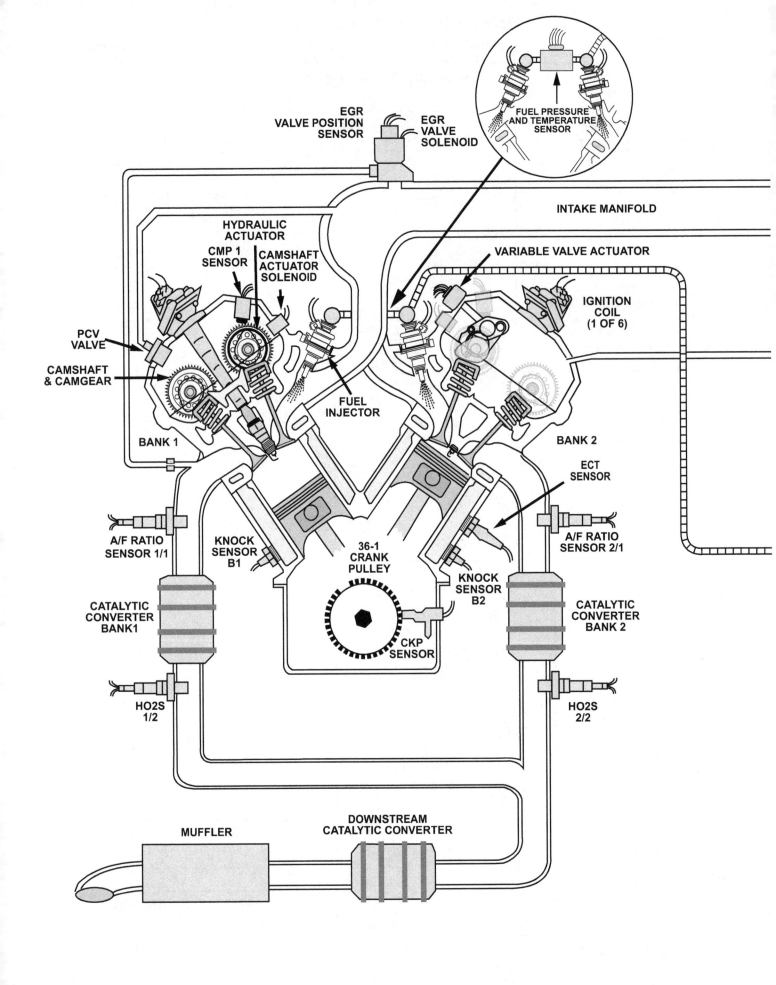

EGR VALVE POSITION SENSOR

EGR VALVE SOLENOID

FUEL PRESSURE AND TEMPERATURE SENSOR

INTAKE MANIFOLD

HYDRAULIC ACTUATOR

CMP 1 SENSOR

CAMSHAFT ACTUATOR SOLENOID

VARIABLE VALVE ACTUATOR

IGNITION COIL (1 OF 6)

PCV VALVE

CAMSHAFT & CAMGEAR

FUEL INJECTOR

BANK 1

BANK 2

ECT SENSOR

A/F RATIO SENSOR 1/1

KNOCK SENSOR B1

36-1 CRANK PULLEY

KNOCK SENSOR B2

A/F RATIO SENSOR 2/1

CATALYTIC CONVERTER BANK1

CATALYTIC CONVERTER BANK 2

CKP SENSOR

HO2S 1/2

HO2S 2/2

MUFFLER

DOWNSTREAM CATALYTIC CONVERTER

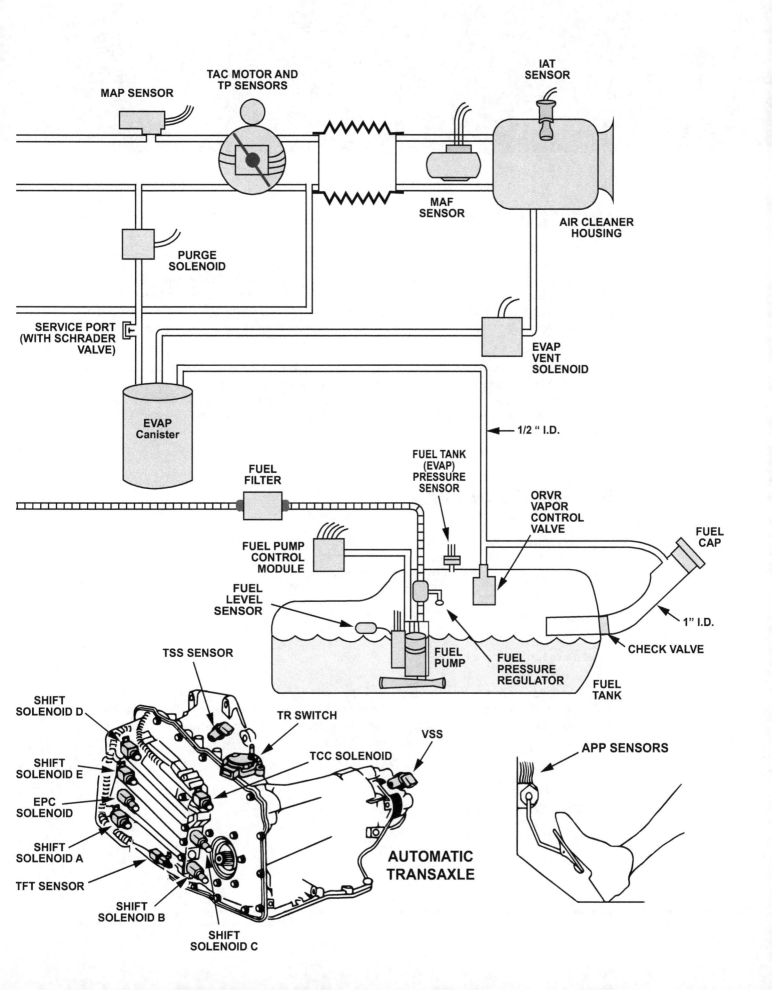

MAP SENSOR

TAC MOTOR AND
TP SENSORS

IAT
SENSOR

MAF
SENSOR

AIR CLEANER
HOUSING

PURGE
SOLENOID

SERVICE PORT
(WITH SCHRADER
VALVE)

EVAP
VENT
SOLENOID

EVAP
Canister

1/2 " I.D.

FUEL
FILTER

FUEL TANK
(EVAP)
PRESSURE
SENSOR

ORVR
VAPOR
CONTROL
VALVE

FUEL
CAP

FUEL PUMP
CONTROL
MODULE

1" I.D.

FUEL
LEVEL
SENSOR

CHECK VALVE

TSS SENSOR

FUEL
PUMP

FUEL
PRESSURE
REGULATOR

FUEL
TANK

SHIFT
SOLENOID D

TR SWITCH

VSS

APP SENSORS

SHIFT
SOLENOID E

TCC SOLENOID

EPC
SOLENOID

SHIFT
SOLENOID A

TFT SENSOR

AUTOMATIC
TRANSAXLE

SHIFT
SOLENOID B

SHIFT
SOLENOID C

Sample Preparation Exams

INTRODUCTION

Included in this section are a series of six individual preparation exams that you can use to help determine your overall readiness to successfully pass the Advanced Engine Performance (L1) ASE certification exam. Located in Section 7 of this book you will find blank answer sheet forms you can use to designate your answers to each of the preparation exams. Using these blank forms will allow you to attempt each of the six individual exams multiple times without risk of viewing your prior responses.

Upon completion of each preparation exam, you can determine your exam score using the answer keys and explanations located in Section 6 of this book. Included in the explanation for each question is the specific task area being assessed by that individual question. This additional reference information may prove useful if you need to refer back to the task list located in Section 4 for additional support.

PREPARATION EXAM 1

1. Technician A says that switching oil viscosities may affect the operation of the variable valve timing (VVT) system. Technician B says that switching oil viscosities may affect the operation of the positive crankcase ventilation (PCV) system. Who is correct?

    A. A only
    B. B only
    C. Both A and B
    D. Neither A nor B

2. An engine misfire cannot be duplicated. Technician A says that spraying a saltwater solution on the spark plug wires may cause the misfire to occur easier. Technician B says that the customer should be quizzed specifically about when the misfire occurs. Who is correct?

    A. A only
    B. B only
    C. Both A and B
    D. Neither A nor B

3. Technician A says that maintaining the proper battery voltage during electronic control module (ECM) reprogramming is critical. Technician B says that if battery voltage falls too low during reprogramming the ECM may have to be replaced. Who is correct?

   A. A only

   B. B only

   C. Both A and B

   D. Neither A nor B

4. A vehicle equipped with a 4-cylinder engine has come into the shop with a diagnostic trouble code (DTC) P0172 – bank 1 rich. Which of the following would LEAST LIKELY cause this problem?

   A. Excessive fuel pressure

   B. A plugged fuel return line

   C. A leaking injector

   D. A plugged fuel filter

5. A vehicle is brought into the shop with a DTC P0171 – System Lean, and a vacuum leak is suspected. Technician A says that a smoke machine can be used to locate the vacuum leak. Technician B says that the smoke should be blown around the intake with the engine running while noting an increase in RPM. Who is correct?

   A. A only

   B. B only

   C. Both A and B

   D. Neither A nor B

6. The tailpipe emissions are being tested to aid in the diagnosis of a poorly running engine with no power on take-off. The readings at idle are as follows:

   HC – 120 ppm

   CO – 0.5 percent

   CO_2 – 12 percent

   O_2 – 10 percent

   What could be causing these readings?

   A. A leaking injector

   B. Low fuel pressure

   C. A restricted air inlet

   D. Stuck open exhaust gas recirculation (EGR) valve

7. The air injection reaction (AIR) system hoses have burned repeatedly on a vehicle. What component should the technician check to prevent this failure again?

   A. AIR pump

   B. AIR diverter valve

   C. Exhaust check valves

   D. Catalytic converter

8. Technician A says that low manifold vacuum can be seen on the scan tool as high manifold absolute power (MAP) sensor voltage readings. Technician B says a clogged exhaust can cause low MAP sensor voltage readings. Who is correct?

    A. A only

    B. B only

    C. Both A and B

    D. Neither A nor B

9. A restricted catalytic converter has been diagnosed. This could be caused by:

    A. A malfunctioning EGR system

    B. A clogged injector

    C. Excessive combustion chamber deposits

    D. Leaking valve guide seals

SCAN TOOL DATA

Engine Coolant Temperature Sensor (ECT) 0.46 Volts	Intake Air Temperature Sensor (IAT) 2.24 Volts	MAP Sensor 1.8 V MAF Sensor 1.1 V	Throttle Position Sensor (TPS) 0.6 Volts
Engine Speed Sensor (RPM) 1500 rpm	Heated O$_2$ HO$_2$S Upstream 0.2–0.5 V Downstream 0.1–0.3 V	Vehicle Speed Sensor (VSS) 0 mph	Battery Voltage (B+) 14.2 Volts
Idle Air Control Valve (IAC) 0 percent	Evaporative Emission Canister Solenoid (EVAP) OFF	Torque Converter Clutch Solenoid (TCC) OFF	EGR Valve Control Solenoid (EGR) 0 percent
Malfunction Indicator Lamp (MIL) OFF	Diagnostic Trouble Codes NONE	Open/Closed Loop CLOSED	Fuel Pump Relay (FP) On
Fuel Level Sensor 3.5 V	Fuel Tank Pressure Sensor 2.0 V	Transmission Fluid Temperature Sensor 0.6 V	Transmission Turbine Shaft Speed Sensor 0 mph
Transmission Range Switch PARK	Trans Pressure Control Solenoid 80%	Transmission Shift Solenoid 1 ON	Transmission Shift Solenoid 2 OFF

| Measured Ignition Timing °BTDC | Base Timing: 10° | Actual Timing: 20° |

10. The scan tool data in the figure above is being reviewed. Technician A says that the data could indicate low fuel pressure. Technician B says that the data could indicate an intake manifold vacuum leak. Who is correct?

    A. A only

    B. B only

    C. Both A and B

    D. Neither A nor B

11. Technician A says that high O_2 emissions levels are an indicator of an efficient running engine. Technician B says that high CO_2 levels are an indicator of a rich condition. Who is correct?

    A. A only

    B. B only

    C. Both A and B

    D. Neither A nor B

12. An oil type air filter can cause contamination of the MAF sensor, which will:

    A. Cause a rich condition

    B. Cause a lean condition

    C. Cause EGR system faults

    D. Cause high CO_2 emissions

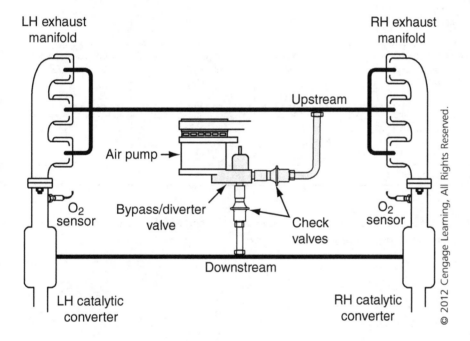

13. The diverter valve in the AIR system pictured in the above figure has failed and continuously pumps air upstream. This will cause:

    A. Low O_2 sensor voltages

    B. High O_2 sensor voltages

    C. Catalyst damage

    D. A rough idle

14. The output of a two wire crank sensor can be tested most accurately by using:

    A. An ohmmeter

    B. An AC voltmeter

    C. A DC voltmeter

    D. A test light

15. Technician A says that a plugged injector can be diagnosed by performing a fuel pressure test. Technician B says that the plugged injector could have been caused by debris in the tank. Who is correct?

    A. A only

    B. B only

    C. Both A and B

    D. Neither A nor B

SCAN TOOL DATA

Engine Coolant Temperature (ECT) Sensor 0.52 volts	Intake Air Temperature (IAT) Sensor 2.4 volts	Manifold Absolute Pressure (MAP) Sensor 1.9 volts	Throttle Position Sensor (TPS) 0.5 volts
Engine Speed Sensor (rpm) 450 rpm	Heated Oxygen Sensor (HO$_2$S) 0.3–0.8 volts	Vehicle Speed Sensor (VSS) 0 mph	Battery Voltage (B+) 12.8 volts
Idle Air Control (IAC) Valve 100 percent	Evaporative Emission Canister Solenoid (EVAP) OFF	Torque Converter Clutch (TCC) Solenoid OFF	EGR Valve Control Solenoid (EGR) 0 percent
Malfunction Indicator Lamp (MIL) OFF	Diagnostic Trouble Codes NONE	Open/Closed Loop CLOSED	Fuel Pump Relay (FP) ON

Measured Ignition Timing °BTDC　　　Base Timing: 10°　　　Actual Timing: 11°

16. A technician is attempting to verify the concern of dying at idle. Based on the above scan tool data, what could be wrong with the vehicle?

    A. Faulty TPS

    B. A charging system fault

    C. Faulty IAC

    D. Faulty MAP sensor

17. An evaporative emissions (EVAP) system leak is being diagnosed, and the system is being smoke tested. Technician A says that the EVAP purge valve will have to be commanded shut since it is normally an open valve. Technician B says that the EVAP vent valve is normally closed and will therefore not have to be commanded shut. Who is correct?

    A. A only

    B. B only

    C. Both A and B

    D. Neither A nor B

18. A vehicle is brought into the shop with an illuminated MIL, a P0305 – cylinder 5 misfire, and a P0302 – cylinder 2 misfire on a V6 engine. The engine uses a waste spark ignition system, and the firing order is 1-2-3-4-5-6. The engine runs smooth with no evidence of a misfire; however, while monitoring the misfire counter, cylinders 2 and 5 show a continuous misfire under all running conditions. This could be caused by:

   A. A damaged crankshaft tone wheel
   B. High resistance in the plug wires
   C. A faulty ECM coil driver
   D. A faulty 2/5 ignition coil

19. An engine with a vacuum leak will have what effect on exhaust emissions?

   A. Lower-than-normal HCs
   B. Lower-than-normal O_2
   C. Higher-than-normal O_2
   D. Higher-than-normal CO_2

20. A vehicle is brought into the shop with an illuminated MIL. The ECM has stored a DTC P0135 – HO_2S heater performance. The voltage drop of the power circuit is 12.1 V. This voltage could indicate:

   A. Low system voltage
   B. An open O_2 sensor heater circuit
   C. Resistance in the power circuit
   D. An open O_2 heater ground circuit

21. Technician A says that the MIL lamp should illuminate for approximately two seconds after the key is turned on. Technician B says that the MIL lamp should illuminate as soon any engine-related DTC is set. Who is correct?

   A. A only
   B. B only
   C. Both A and B
   D. Neither A nor B

22. A vehicle fails a loaded I/M test for excessive HC emissions. This could be caused by:

   A. Leaking vacuum hose
   B. Fouled spark plug
   C. Cracked spark plug
   D. Hung open EVAP purge valve

23. A slipped timing belt is suspected on a poorly running, non-interference engine. What vehicle data would support this diagnosis?

   A. Low O_2 sensor voltage
   B. High MAP sensor voltage
   C. A faulty cam sensor signal
   D. High MAF sensor signal

24. A vehicle starts, runs for two seconds and dies. The LEAST LIKELY cause of this condition could be:

    A.   An immobilizer system fault

    B.   A faulty IAC valve

    C.   Low fuel pressure

    D.   A dirty throttle body

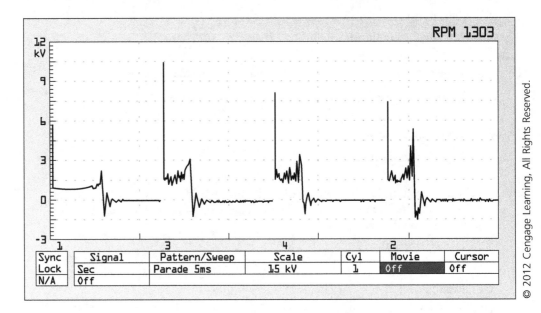

25. The secondary voltage patterns shown in the above figure were taken after performing a tune-up. The waveform could be caused by:

    A.   A rich condition

    B.   A bridged #3 plug gap

    C.   Excessive plug wire resistance

    D.   A too-small #1 plug gap

26. An ECM has been replaced several times for a non-functioning injector and a DTC P0201 – injector #1 control circuit. The injector will function for a while with the new ECM, but will soon fail again. Which test could find the cause of this failure?

    A.   Injector resistance

    B.   Injector voltage drop

    C.   Control circuit resistance

    D.   Power circuit voltage drop

27. Technician A says that a poorly maintained cooling system can cause excessive NO_x formation. Technician B says that plugged EGR passages can also cause excessive NO_x formation. Who is correct?

    A.   A only

    B.   B only

    C.   Both A and B

    D.   Neither A nor B

28. Fuel pressure is being monitored on a fuel pressure gauge. Technician A says that the pressure should go down when the vacuum line is removed from the fuel pressure regulator. Technician B says that fuel should drain from the regulator when the vacuum line is removed. Who is correct?

    A. A only
    B. B only
    C. Both A and B
    D. Neither A nor B

29. Which piece of scan tool data would be the MOST helpful when trying to diagnose a plugged catalytic converter?

    A. RPM
    B. MAP
    C. HO_2S
    D. TPS

30. The O_2 sensor is reading low voltage (below 250 mV). What exhaust gas readings can be expected?

    A. High O_2 content
    B. High CO_2 content
    C. Low NO_x content
    D. Low O_2 content

31. The current draw of a fuel pump is lower than specified. This could be caused by:

    A. Shorted fuel pump windings
    B. A stuck closed pressure regulator
    C. A clogged fuel filter
    D. Excessive circuit resistance

32. During a plug-in emissions test, the catalyst monitor displays "Not Complete." What is the possible cause?

    A. A failed catalytic converter
    B. The ECM memory has been reset
    C. The engine is not up to operating temperature
    D. An exhaust manifold leak

33. A malfunction of the PCV system is likely to cause all of these problems EXCEPT:

    A. Excessive blowby
    B. An illuminated MIL
    C. Rough/unstable idle speed
    D. Engine oil leaks

34. Positive fuel trim numbers can be caused by:

    A. High fuel pressure
    B. Clogged fuel filter
    C. Stuck closed fuel pressure regulator
    D. An injector hung open

35. While being tested by a scan tool, TPS voltage drops to near zero (0.02 V). This indicates:

    A. The throttle is at idle position
    B. A bad spot in the TPS
    C. The scan tool is processing data
    D. Normal activity

36. A vehicle has failed an emissions test with high NO_x levels, and tampering is suspected. The emission system that was most likely tampered with would be:

    A. EVAP system
    B. PCV system
    C. EGR system
    D. ORVR system

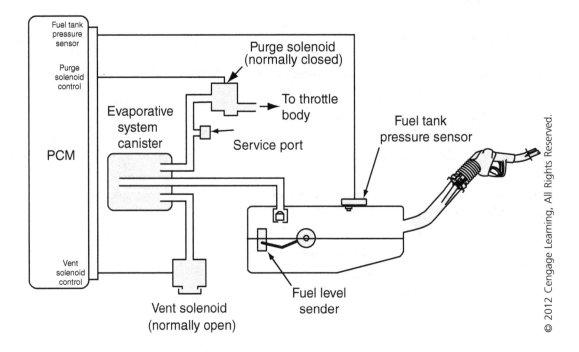

37. The customer has a concern of a slow fill when refueling the fuel system in the above figure, and the pump nozzle keeps kicking off. This could be caused by:

    A. A stuck closed vent solenoid
    B. A leaking hose from the tank to the canister
    C. A leaking purge solenoid
    D. A stuck closed purge solenoid

38. A vehicle is brought into the shop with a rough idle and both upstream O_2 sensors displaying high voltage. Which of the following could be the cause?

    A. Faulty O_2 sensors

    B. An intake manifold vacuum leak

    C. Restricted air filter

    D. Stuck open EGR valve

39. A vehicle fails a loaded emissions test for excessive CO. This could be caused by:

    A. An intake manifold vacuum leak

    B. Low fuel pump output

    C. Clogged fuel injectors

    D. Plugged fuel return line

40. All of the following emission system faults could cause a rough idle EXCEPT:

    A. A stuck open EGR valve

    B. A stuck open EVAP vent valve

    C. A stuck open EVAP purge valve

    D. A leaking PCV vacuum hose

41. An improperly maintained cooling system is likely to cause what problem?

    A. Poor fuel economy

    B. Rough idle

    C. Detonation

    D. Low power output

42. All of the following can cause excessive CO tailpipe emissions EXCEPT:

    A. A leaking injector

    B. A leaking vacuum hose

    C. Excessive fuel pressure

    D. A hanging EVAP purge solenoid

43. The EVAP system for the composite vehicle is being diagnosed. Technician A says that the EVAP system monitor will run no matter what the level of fuel in the tank. Technician B says that the EVAP system monitor will run no matter what the engine temperature. Who is correct?

    A. A only

    B. B only

    C. Both A and B

    D. Neither A nor B

44. The composite vehicle's VVT system:

    A. Changes the timing of the intake cam
    B. Changes the timing of the exhaust cam
    C. Changes the timing of both cams
    D. Changes the cam's duration

45. The composite vehicle fails a loaded I/M test for excessive NO_x emissions. The possible cause for this condition could be:

    A. A ruptured EGR vacuum line
    B. A bad connection at ECM terminal 108
    C. A bad connection at ECM terminal 28
    D. A leaking EGR valve diaphragm

46. The composite vehicle is brought into the shop with a DTC P0172 – Fuel system bank 1 rich, and DTC P0174 – Fuel system bank 2 rich. What could cause the condition?

    A. A kinked fuel return hose
    B. A stuck open injector
    C. A cracked MAP sensor vacuum hose
    D. Excessive fuel pressure

47. The composite vehicle enters the shop with a DTC P0341 – camshaft position (CMP) sensor performance. Technician A says that this could be caused by an open CMP sensor. Technician B says that this could be caused by excessive camshaft end-play. Who is correct?

    A. A only
    B. B only
    C. Both A and B
    D. Neither A nor B

48. After sitting for ten minutes, the composite vehicle is hard to start, and fuel pressure measures 32 psi (220 kPa). Once started, the engine runs fine with no DTCs. Which of the following could be the cause?

    A. Leaking fuel pump check valve
    B. Low fuel pump output
    C. Leaking injector o-ring
    D. A lost cam sensor signal

49. The composite vehicle is being diagnosed for a P0302 – cylinder 2 misfire. The #2 ignition coil secondary resistance is 2 KΩ. This indicates:

    A. A shorted secondary coil
    B. An open secondary coil
    C. A shorted primary coil
    D. A normal coil

SCAN TOOL DATA			
Engine Coolant Temperature (ECT) Sensor 4.3 volts	Intake Air Temperature (IAT) Sensor 2.4 volts	Manifold Absolute Pressure (MAP) Sensor 1.8 volts	Throttle Position Sensor (TPS) 2.4 volts
Engine Speed Sensor (rpm) 1700 rpm	Heated Oxygen Sensor (HO_2S) 0.9 volts	Vehicle Speed Sensor (VSS) 55 mph	Battery Voltage (B+) 14.3 volts
	Evaporative Emission Canister Solenoid (EVAP) ON	Torque Converter Clutch (TCC) Solenoid ON	EGR Valve Control Solenoid (EGR) 30 percent
Malfunction Indicator Lamp (MIL) ON	Diagnostic Trouble Codes NONE	Open/Closed Loop CLOSED	Fuel Pump Relay (FP) ON

Measured Ignition Timing °BTDC	Base Timing: 10°	Actual Timing: 11°

50. The composite vehicle is brought into the shop with the concern of black smoke from the tailpipe. Use the data displayed in the above figure to diagnose the fault.

    A. A faulty O_2 sensor

    B. A skewed ECT sensor

    C. A skewed IAT sensor

    D. A stuck open EVAP purge valve

PREPARATION EXAM 2

1. Which of the following emissions changes is LEAST LIKELY to be caused by a clogged fuel injector?

    A. Decreased CO_2 emissions

    B. Decreased O_2 emissions

    C. Decreased NO_x emissions

    D. Increased CO emissions

2. A vehicle was repaired for a DTC P0171 – bank 1 lean. Technician A says that fuel trim percentages can be used to verify the repair. Technician B says that monitoring the O_2 sensor voltages will verify the repair. Who is correct?

    A. A only

    B. B only

    C. Both A and B

    D. Neither A nor B

3. A resistor has been installed in series with the intake air temperature (IAT) sensor in an attempt to increase engine output. This will have what affect on the IAT reading?

    A. Decreased fuel delivery

    B. Decreased signal voltage

    C. Increased temperature reading

    D. Increased signal voltage

4. A technician is testing the integrity of the EVAP system after a major fuel system repair. With the smoke tester connected and turned on, the indicator pellet in the flow gauge remains at the top of the flow gauge. This indicates:

    A. A large leak in the system

    B. A small leak in the system

    C. No leak in the system

    D. Nothing

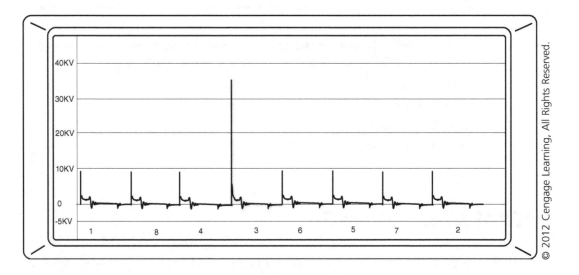

5. The pattern in the above figure is displayed while testing secondary voltage. This condition will cause:

    A. Low O_2 emissions

    B. Low NO_x emissions

    C. Excessive HC emissions

    D. Excessive CO emissions

6. Black smoke is coming from the exhaust under all engine conditions. Technician A says that a throttle position sensor (TPS) that is fixed around 4.5 volts could cause this. Technician B says that an engine coolant temperature (ECT) sensor that is fixed around 4.5 volts could also cause this. Who is correct?

    A. A only

    B. B only

    C. Both A and B

    D. Neither A nor B

7. Technician A says that spark knock can be caused by excessive scale in the cooling system. Technician B says that spark knock can also be caused by using fuel with a too-high octane rating. Who is correct?

    A. A only

    B. B only

    C. Both A and B

    D. Neither A nor B

8. A vehicle is brought into the shop with a poor fuel economy concern. Technician A says that installing larger wheels and tires can throw the odometer off and cause this concern. Technician B says that installing a higher-ratio final drive gearset can cause this concern. Who is correct?

    A. A only

    B. B only

    C. Both A and B

    D. Neither A nor B

9. An O_2 sensor is fixed lean. Which of the following can be performed to change O_2 voltage and confirm the operation of the sensor?

    A. Create a vacuum leak

    B. Unplug an injector

    C. Remove spark from a cylinder

    D. Open the EVAP purge valve

10. The EGR system is being tested after cleaning clogged passages. What should result when the valve is commanded open at idle with the scan tool?

    A. Rough unstable idle

    B. Increase in idle speed

    C. Black smoke from the exhaust

    D. Low O_2 sensor voltage

11. A vehicle has come into the shop with an intermittent DTC P0340 – camshaft sensor circuit malfunction. Technician A says that this may be caused by a faulty cam sensor. Technician B says that the cam sensor can be damaged by excessive cam thrust. Who is correct?

    A. A only

    B. B only

    C. Both A and B

    D. Neither A nor B

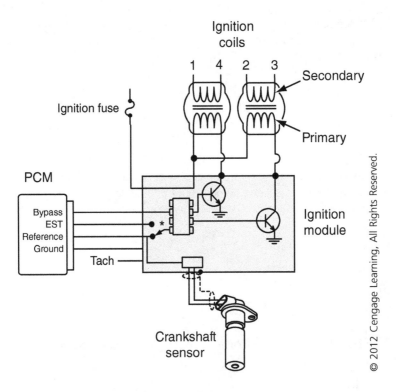

12. Technician A says that the ignition module in the above image controls the current through the primary ignition coils. Technician B says that the module must have a cam sensor input so that it knows when to turn on each coil. Who is correct?

    A. A only
    B. B only
    C. Both A and B
    D. Neither A nor B

13. The throttle of a vehicle with throttle actuator control (TAC) is unresponsive, and idle speed is high. Technician A says that there may not be any power to the throttle motor. Technician B says that one faulty TPS could cause this condition. Who is correct?

    A. A only
    B. B only
    C. Both A and B
    D. Neither A nor B

14. Technician A says that the fuel level sensor must be functional before the EVAP system can be tested. Technician B says that the O_2 sensor monitor must run and pass before the catalytic converter monitor will run. Who is correct?

    A. A only
    B. B only
    C. Both A and B
    D. Neither A nor B

15. A monitor readiness status of "Complete" indicates:

    A. The vehicle will pass an I/M test
    B. The monitor has passed
    C. The monitor has failed
    D. The monitor has run

16. Exhaust emissions are being tested on a vehicle. O_2 levels are higher than normal, and CO_2 levels are low. Which of the following would LEAST LIKELY cause this condition?

    A. A restricted injector
    B. A stuck closed PCV valve
    C. A clogged fuel filter
    D. A vacuum leak

17. Technician A says that it is not possible to have both high O_2 and CO emissions. Technician B says that O_2 is a lean indicator and CO is a rich indicator. Who is correct?

    A. A only
    B. B only
    C. Both A and B
    D. Neither A nor B

18. What effect would a stuck open EGR valve have on O_2 sensor readings?

    A. Low steady voltage readings
    B. Slowly switching voltage readings
    C. Rapidly switching voltage readings
    D. High steady voltage readings

19. A vehicle with a return type fuel system is hard to start after sitting and has black smoke from the exhaust after finally starting. This could be caused by:

    A. A leaking fuel pressure regulator
    B. Low fuel pump output
    C. A leaking fuel pump check valve
    D. A plugged fuel filter

20. The ignition module on a distributor ignition (DI) vehicle has been replaced several times after it has failed. Technician A says that this can be caused by a shorted magnetic pick-up. Technician B says that this can also be caused by a shorted ignition coil. Who is correct?

    A. A only
    B. B only
    C. Both A and B
    D. Neither A nor B

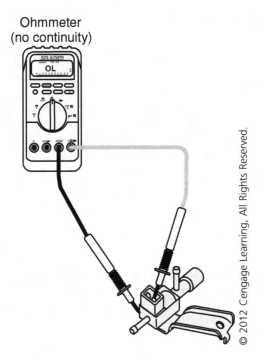

Ohmmeter
(no continuity)

© 2012 Cengage Learning, All Rights Reserved.

21. The EVAP purge solenoid in the above figure is being tested. Technician A says that the solenoid is open. Technician B says that the solenoid must be replaced. Who is correct?

    A. A only

    B. B only

    C. Both A and B

    D. Neither A nor B

22. A vehicle has an illuminated MIL and a DTC P0123 – TPS high voltage. Technician A says that an open signal return wire could cause this code. Technician B says that a shorted TPS could cause this DTC. Who is correct?

    A. A only

    B. B only

    C. Both A and B

    D. Neither A nor B

23. A vehicle has failed a no-load I/M test for high HC emissions, but HC emissions are acceptable at higher RPM. Technician A says that this can be caused by a fouled spark plug. Technician B says that this could be caused by low fuel pressure. Who is correct?

    A. A only

    B. B only

    C. Both A and B

    D. Neither A nor B

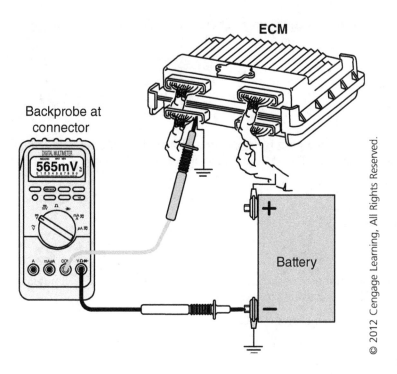

ECM

Backprobe at connector

565mV

Battery

© 2012 Cengage Learning, All Rights Reserved.

24. A vehicle is brought into the shop with multiple DTCs and numerous drivability concerns. Testing of one of the ECM grounds is displayed above. This test indicates:

   A. High resistance in the ECM power circuit

   B. High resistance in the ECM ground circuit

   C. Continuity of the ground circuit

   D. The ground circuit is shorted to power

25. A cracked exhaust manifold will cause O_2 sensor readings to:

   A. Read low voltage

   B. Read high voltage

   C. Switch rapidly

   D. Be fixed at 450 mV

Cylinder	1	2	3	4	5	6
High reading	225	225	225	225	225	225
Low reading	100	100	100	100	115	115
Amount of drop	125	125	125	125	110	110

26. The results from an injector balance test are displayed in the table above. These results can cause:

   A. A rich condition

   B. An engine miss

   C. A lean condition

   D. Low engine power

27. An air injection system that constantly pumps air upstream will cause the O_2 sensors to:

    A. Read rich
    B. Create low voltage
    C. Create high voltage
    D. Fluctuate rapidly

28. An ECT sensor is being checked for proper operation. While the voltage is being monitored on the scan tool after a cold start, what should the voltage do as the sensor heats up?

    A. Switch rapidly
    B. Steadily increase
    C. Steadily decrease
    D. Slowly fluctuate

29. The scan tool will not communicate with the vehicle's ECM, and the MIL will not illuminate during the bulb test. The most likely cause of this is:

    A. A bad diagnostic link connection (DLC)
    B. No power to the DLC
    C. A malfunctioning scan tool
    D. An open ECM ground

30. A vehicle has a high idle and no A/C. Which of the following could cause both of these symptoms?

    A. A shorted power steering pressure switch
    B. An intake manifold vacuum leak
    C. A hanging throttle plate
    D. Higher-than-normal TPS voltage

31. The outlet temperature of the catalytic converter is 910°F (488°C) and the inlet is 1005°F (541°C). This indicates:

    A. The converter is overheating
    B. There is a misfire
    C. A degraded converter
    D. Normal operation

32. Which of the following valves, if stuck open, would cause highest increase in CO emissions?

    A. EGR
    B. EVAP vent
    C. EVAP purge
    D. Diverter

33. A vehicle arrives in the shop with a DTC P0401 – EGR insufficient flow. All of the following could cause this EXCEPT:

    A. Aftermarket exhaust system

    B. Restricted catalytic converter

    C. Broken EGR valve vacuum hose

    D. Plugged EGR passages

34. A stuck open EVAP purge solenoid would cause:

    A. Increased CO emissions

    B. Increased CO_2 emissions

    C. Decreased NO_x emissions

    D. Decreased HC emissions

35. A vehicle with a return type fuel system has failed an emissions test with the following readings at idle:

    HC – 20 ppm
    CO – 3 percent
    CO_2 – 10 percent
    O_2 – 0 percent

    Which of the following could cause these readings?

    A. An intake manifold vacuum leak

    B. A plugged fuel filter

    C. A leaking fuel pressure regulator

    D. A plugged EGR passage

36. Technician A says that most immobilizer systems will turn off the fuel injectors if there is a fault in the anti-theft system. Technician B says that the ignition system is also disabled. Who is correct?

    A. A only

    B. B only

    C. Both A and B

    D. Neither A nor B

37. A fuel injector can be tested for all of the following EXCEPT:

    A. Flow

    B. Resistance

    C. Pressure

    D. Current draw

38. Engine detonation can be caused by:

    A. An inoperative EVAP purge valve

    B. A clogged catalytic converter

    C. A clogged PCV valve

    D. An inoperative EGR valve

39. The composite vehicle is brought into the shop with a DTC P0302 – cylinder #2 misfire. The cause of this misfire could be:

    A. A blown fuse #4

    B. An open wire from the coil to terminal 17 of the ECM

    C. An open wire from the coil to terminal 18 of the ECM

    D. High secondary plug wire resistance

40. The ECM has been replaced several times because the A/C clutch relay driver has failed. This could be caused by:

    A. A shorted relay coil

    B. High relay coil resistance

    C. High A/C clutch coil resistance

    D. A shorted A/C clutch coil

41. The composite vehicle is slow to fill when refueling. This could be due to all of the following EXCEPT:

    A. A pinched hose from the vent solenoid to the intake

    B. A stuck closed purge solenoid

    C. A pinched hose from the tank to the EVAP canister

    D. A stuck closed vent solenoid

42. The composite vehicle engine speed will not rise above 6000 rpm. This is most likely due to:

    A. ECM programming

    B. A restricted catalytic converter

    C. Low fuel pump output pressure

    D. A slipped timing chain

43. The composite vehicle has been brought into the shop with the MIL illuminated, a DTC P0172 – bank 1 rich, and P0174 – bank 2 rich. The cause for this condition could be:

    A. A leaking fuel injector

    B. A leaking fuel pressure regulator

    C. A short to ground at ECM terminal 110

    D. A short to ground at ECM terminal 120

SCAN TOOL DATA

Engine Coolant Temperature Sensor (ECT) 0.35 Volts	Intake Air Temperature Sensor (IAT) 4.50 Volts	MAP Sensor 4.5 V MAF Sensor 0.2 V	Throttle Position Sensor (TPS) 0.6 Volts
Engine Speed Sensor (RPM) 0 rpm	Heated O_2 HO_2S Upstream 0 V Downstream 0 V	Vehicle Speed Sensor (VSS) 0 mph	Battery Voltage (B+) 12.6 Volts
	Evaporative Emission Canister Solenoid (EVAP) OFF	Torque Converter Clutch Solenoid (TCC) OFF	EGR Valve Control Solenoid (EGR) 0 percent
Malfunction Indicator Lamp (MIL) ON	Diagnostic Trouble Codes NONE	Open/Closed Loop OPEN	Fuel Pump Relay (FP) OFF
Fuel Level Sensor 3.5 V	Fuel Tank Pressure Sensor 2.5 V	Transmission Fluid Temperature Sensor 3.5 V	Transmission Turbine Shaft Speed Sensor 0 mph
Transmission Range Switch PARK	Trans Pressure Control Solenoid 0	Transmission Shift Solenoid 1 ON	Transmission Shift Solenoid 2 OFF

Measured Ignition Timing °BTDC   Base Timing: 10°   Actual Timing: 20°

44. The composite vehicle has been brought into the shop because of poor fuel economy. Based on the key on and the engine off (KOEO) scan tool readings shown in the figure above, what could be the cause?

    A. A faulty ECT sensor

    B. A cracked MAP sensor vacuum line

    C. A stuck closed EGR valve

    D. A faulty IAT sensor

45. The composite vehicle will not start. What is the minimum voltage required while measuring the output of the crankshaft position sensor while cranking?

    A. 5.0 VAC

    B. 2.0 VDC

    C. 0.5 VDC

    D. 0.2 VAC

46. A DTC P0102 – MAF sensor low voltage could be caused by all of the following EXCEPT:

    A. A clogged exhaust

    B. Air duct leaks

    C. Water intrusion

    D. A vacuum leak

47. The composite vehicle is brought into the shop with a spark knock concern. What should the technician do?

   A. Adjust ignition timing
   B. Fill the tank with quality fuel
   C. Check for excessive fuel pressure
   D. Adjust the TPS

48. The composite vehicle is brought into the shop with a DTC P0031 - AFR Sensor 1/1 heater circuit. Technician A says that this could be caused by a blown fuse #20. Technician B says that this could be caused by an open in the wire between terminal B of the AFR sensor and terminal 151 of the ECM. Who is correct?

   A. A only
   B. B only
   C. Both A and B
   D. Neither A nor B

49. The composite vehicle has a rough idle but drives normally at cruise. Technician A says that terminal 108 at the ECM could be grounded. Technician B says that a grounded ECM terminal 104 can also cause this symptom. Who is correct?

   A. A only
   B. B only
   C. Both A and B
   D. Neither A nor B

50. The composite vehicle has failed a plug in emissions test because the catalytic converter monitor displayed "Not Complete." Technician A says that this means the monitor has not run. Technician B says that the monitor may not have run because one of the AFR sensor monitors may have failed. Who is correct?

   A. A only
   B. B only
   C. Both A and B
   D. Neither A nor B

PREPARATION EXAM 3

1. Technician A says that excessive combustion chamber deposits could cause high NO_x emissions. Technician B says that excessive combustion chamber deposits could also cause excessive CO_2 emissions. Who is correct?

   A. A only
   B. B only
   C. Both A and B
   D. Neither A nor B

2. While testing the function of the exhaust gas recirculation (EGR) system, what should be noticed when the EGR valve is opened with the key on and the engine running (KOER)?

   A. Increased engine speed
   B. Black exhaust smoke
   C. Lowered O_2 sensor readings
   D. Rough or unstable idle

3. A throttle position sensor (TPS) is suspected of causing a hesitation. Which of the following tools would LEAST LIKELY be used to test the TPS?

   A. Test light
   B. Voltmeter
   C. Oscilloscope
   D. Ohmmeter

4. Technician A says that a misfiring cylinder due to a fouled spark plug will increase HC emissions. Technician B says that the fouled plug will increase O_2 levels. Who is correct?

   A. A only
   B. B only
   C. Both A and B
   D. Neither A nor B

5. The main sensor used for fuel control when the vehicle is in closed loop is:

   A. TPS
   B. O_2
   C. ECT
   D. MAP

6. Technician A says that electronic control module (ECM) reprogramming can be done to correct a problem when instructed by a TSB. Technician B says ECM reprogramming can be done during routine maintenance to prevent problems. Who is correct?

   A. A only
   B. B only
   C. Both A and B
   D. Neither A nor B

7. Technician A says a cracked spark plug will cause firing voltage to increase. Technician B says that a fouled spark plug will also cause firing voltage to increase. Who is correct?

   A. A only
   B. B only
   C. Both A and B
   D. Neither A nor B

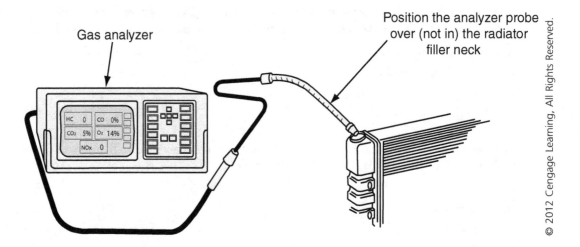

Gas analyzer

Position the analyzer probe over (not in) the radiator filler neck

8. What procedure is being done in the above picture?

A. Testing engine coolant level

B. Measuring coolant freeze protection

C. Checking for a blown head gasket

D. Calibrating the exhaust gas analyzer

9. A vehicle with a return-type fuel system idles roughly and blows black smoke after a hot soak. Technician A says that this could be caused by low fuel pressure. Technician B says that a leaking fuel pressure regulator can cause this condition. Who is correct?

A. A only

B. B only

C. Both A and B

D. Neither A nor B

10. What condition should be tested for when O_2 levels are high and CO_2 levels are low?

A. A lean condition

B. A rich condition

C. Low idle speed

D. High fuel pressure

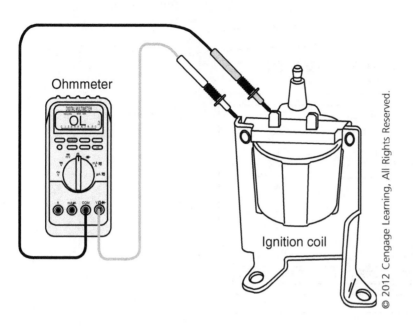

Ignition coil

© 2012 Cengage Learning, All Rights Reserved.

11. What is being tested in the above picture?

   A. Secondary coil resistance

   B. Primary coil resistance

   C. Secondary coil voltage

   D. Primary coil short

12. The ECM has been replaced on a vehicle. Technician A says that the anti-theft system may have to be reprogrammed to recognize the coded ignition key. Technician B says that it is only necessary to program one key. Who is correct?

   A. A only

   B. B only

   C. Both A and B

   D. Neither A nor B

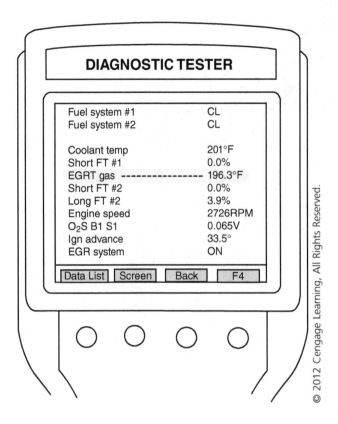

DIAGNOSTIC TESTER

Fuel system #1	CL
Fuel system #2	CL
Coolant temp	201°F
Short FT #1	0.0%
EGRT gas ----------------	196.3°F
Short FT #2	0.0%
Long FT #2	3.9%
Engine speed	2726RPM
O_2S B1 S1	0.065V
Ign advance	33.5°
EGR system	ON

Data List	Screen	Back	F4

13. The EGR system is being tested on a vehicle. The above data is displayed when the EGR valve is actuated with the scan tool. This is an indication of:

   A. EGR flow to one bank only

   B. Too much EGR flow

   C. Insufficient EGR flow

   D. An overheated EGR valve

14. An ECM has been diagnosed with a faulty injector driver. This could be caused by:

   A. Low circuit current

   B. High circuit resistance

   C. Low circuit resistance

   D. Low circuit voltage

15. A vehicle has been brought into the shop several times for a failed camshaft position (CMP) sensor. The root cause of this problem could be:

   A. Excessive cam thrust

   B. A slipped timing belt

   C. Improper engine oil

   D. Excessive timing belt tension

16. Spark knock can be caused by which of the following?

    A. Colder spark plugs than specified

    B. Hotter spark plugs than specified

    C. Smaller plug gap than specified

    D. Larger plug gap than specified

17. Which of the following would LEAST LIKELY cause a high percentage IAC command by the ECM?

    A. A dirty throttle plate

    B. Dirty IAC passages

    C. A faulty IAC valve

    D. A vacuum leak

18. A failed catalytic converter would cause lower output of which exhaust gas?

    A. NO_x

    B. CO_2

    C. HC

    D. O_2

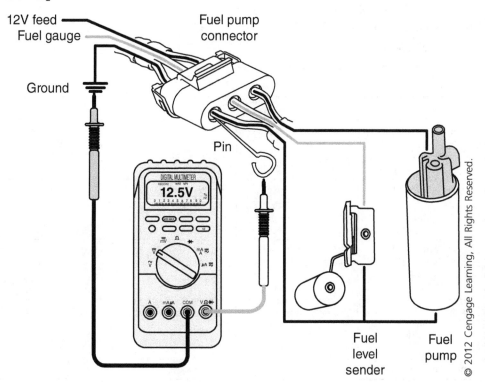

19. A suspected faulty fuel pump was replaced and did not correct the concern. The meter above displays 12.5 V. Technician A says that this reading is normal when the fuel pump is not commanded on. Technician B says that this could indicate a faulty fuel pump if the pump is being commanded on. Who is correct?

    A. A only

    B. B only

    C. Both A and B

    D. Neither A nor B

20. A vehicle starts and dies. Technician A says that a malfunction in the anti-theft system can cause these symptoms. Technician B says that a malfunctioning IAC valve can cause this condition. Who is correct?

    A. A only
    B. B only
    C. Both A and B
    D. Neither A nor B

21. A vehicle failed the initial idle exhaust emissions test, but passed the second-chance test. Technician A says that the vehicle was not preconditioned before the test. Technician B says that the catalytic converter may not have been hot enough. Who is correct?

    A. A only
    B. B only
    C. Both A and B
    D. Neither A nor B

22. A vehicle was brought into the shop as a no start. It was found that the distributor rotor had a hole in it. The most likely cause for this is:

    A. Too little spark plug gap
    B. A shorted ignition coil
    C. High coil wire resistance
    D. High secondary wire resistance

SCAN TOOL DATA			
Engine Coolant Temperature Sensor (ECT) 0.42 Volts	Intake Air Temperature Sensor (IAT) 2.24 Volts	Manifold Absolute Pressure Sensor (MAP) 2.75 Volts	Throttle Position Sensor (TPS) 0.6 Volts
Engine Speed Sensor (RPM) 750 rpm	Heated Oxygen Sensor (HO$_2$S) 0.3–0.7 Volts	Vehicle Speed Sensor (VSS) 0 mph	Battery Voltage (B+) 14.2 Volts
Idle Air Control Valve (IAC) 30 percent	Evaporative Emission Canister Solenoid (EVAP) OFF	Torque Converter Clutch Solenoid (TCC) OFF	EGR Valve Control Solenoid (EGR) 0 percent
Malfunction Indicator Lamp (MIL) OFF	Diagnostic Trouble Codes NONE	Open/Closed Loop CLOSED	Fuel Pump Relay (FP) ON

Measured Ignition Timing °BTDC   Base Timing: 10°   Actual Timing: 11°

23. Technician A says that a lean operating engine can cause the data that is shown in the above figure. Technician B says that the data shown can be caused by a plugged catalytic converter. Who is correct?

    A. A only
    B. B only
    C. Both A and B
    D. Neither A nor B

24. A vehicle owner has installed a colder-than-specified thermostat in the vehicle. This will affect vehicle emissions by:

    A. Increasing HC emissions
    B. Increasing NO_x emissions
    C. Increasing CO_2 emissions
    D. Decreasing CO emissions

25. Measured current draw on a fuel pump circuit is lower than normal. This could be caused by:

    A. A short to ground
    B. An open fuel pump wire
    C. An open relay coil
    D. A corroded connector

26. Positive fuel trim numbers could be caused by:

    A. Excessive fuel pressure
    B. A clogged fuel filter
    C. A leaking injector
    D. A leaking purge valve

27. Technician A says that many states are not equipped to perform the dyno test on all-wheel drive vehicles. Technician B says that all-wheel drive vehicles are usually exempt from emissions testing because they cannot be dyno tested. Who is correct?

    A. A only
    B. B only
    C. Both A and B
    D. Neither A nor B

28. The o-ring seal on the gas cap has many small cracks. Technician A says that these cracks can cause the MIL to illuminate and an EVAP DTC to set if the leak is larger than 0.020" (0.508 mm). Technician B says that the cracks can cause the vehicle to fail an I/M test. Who is correct?

    A. A only
    B. B only
    C. Both A and B
    D. Neither A nor B

29. A vehicle entered the shop with a DTC. The vehicle was diagnosed and repaired, and the DTC was cleared. To verify the repair, the technician should:

    A. Test drive the vehicle while monitoring scan tool data
    B. Refer to the DTC enabling criteria and test drive
    C. Perform a functional test of the engine control system
    D. Inform the customer to return for a follow-up inspection

30. A vehicle fails a loaded I/M emissions test for high HC emissions. HC emissions were acceptable at idle, but increased under a load. This could be caused by:

    A. A leaking fuel injector
    B. A vacuum leak
    C. Excessive fuel pressure
    D. Excessive plug wire resistance

31. While a vehicle with an on-board refueling vapor recovery (ORVR) system is being refueled, fuel vapors are:

    A. Vented to the air intake
    B. Stored in the fuel tank
    C. Stored in the EVAP canister
    D. Vented to the atmosphere

32. A lower-than-normal MAF sensor reading can be caused by all of the following EXCEPT:

    A. A vacuum leak
    B. A restricted exhaust
    C. MAF contamination
    D. High fuel pressure

33. A customer is concerned that water is dripping out of the vehicle's tailpipe. Technician A says that this could be caused by a malfunctioning catalytic converter. Technician B says that the vehicle is running too lean. Who is correct?

    A. A only
    B. B only
    C. Both A and B
    D. Neither A nor B

34. To best find out what conditions were present when a DTC set in the vehicle's ECM, the technician should:

    A. Refer to freeze frame data
    B. Refer to the DTC enabling criteria
    C. Question the customer
    D. Clear the code and attempt to reset it

35. A vehicle has failed an I/M test for high CO emissions. This could be caused by:

    A. Overfilling the fuel tank
    B. Low fuel system pressure
    C. An intake manifold vacuum leak
    D. An inoperative EGR valve

36. All of the following could increase HC exhaust emissions EXCEPT:

    A. A vacuum leak
    B. Stuck open thermostat
    C. Burned exhaust valve
    D. Clogged EGR passage

37. What could cause a vehicle to fail the two-speed idle test for excessive CO emissions?

    A. A leaking vacuum controlled fuel pressure regulator
    B. A leaking PCV system vacuum hose
    C. A restricted fuel filter
    D. A restricted catalytic converter

38. The long-term fuel trim obtained from a vehicle's scan tool data is -20 percent. This data indicates:

    A. The ECM is correcting for a lean condition

    B. The ECM is correcting for a rich condition

    C. A vacuum leak could be present

    D. The ECM is adding more fuel than normal

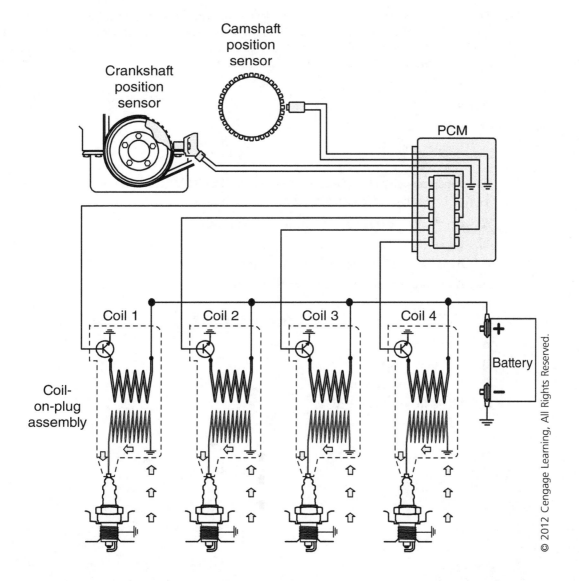

39. A vehicle with the ignition system shown in the above figure will not start and is not sparking on any cylinder. All of the following could cause this EXCEPT:

    A. A faulty cam sensor

    B. A faulty coil

    C. A faulty crank sensor

    D. A faulty ECM

40. A DTC P0401 – EGR insufficient flow has set on the composite vehicle. The EGR solenoid winding resistance is 10 KΩ. This reading indicates:

    A. A normal resistance reading

    B. A shorted EGR solenoid

    C. An open EGR solenoid

    D. Excessive EGR solenoid resistance

41. The composite vehicle's VVT system has failed, and the ECM cannot make any camshaft timing adjustments. Technician A says that no oil pressure to the actuator control solenoid can cause this condition. Technician B says that the cams will default to the fully advanced position when the VVT system fails. Who is correct?

    A. A only

    B. B only

    C. Both A and B

    D. Neither A nor B

42. The composite vehicle's throttle is unresponsive. The RPM is at 1400, and the throttle is fixed at 15 percent. This could be caused by:

    A. One faulty accelerator pedal position (APP) sensor

    B. One faulty TPS

    C. A malfunctioning throttle actuator motor

    D. A throttle actuator that needs to be relearned

43. A failure in the PCV system could cause all of the following conditions EXCEPT:

    A. Excessive oil leaks

    B. Rough or unstable idle

    C. Increased fuel economy

    D. DTCs to be set in the ECM

SCAN TOOL DATA

Engine Coolant Temperature Sensor (ECT) 0.35 Volts	Intake Air Temperature Sensor (IAT) 4.50 Volts	MAP Sensor 4.5 V MAF Sensor 0.2 V	Throttle Position Sensor (TPS) 0.6 Volts
Engine Speed Sensor (RPM) 0 rpm	Heated O₂ HO₂S Upstream 0 V Downstream 0 V	Vehicle Speed Sensor (VSS) 0 mph	Battery Voltage (B+) 12.6 Volts
	Evaporative Emission Canister Solenoid (EVAP) OFF	Torque Converter Clutch Solenoid (TCC) OFF	EGR Valve Control Solenoid (EGR) 0 percent
Malfunction Indicator Lamp (MIL) ON	Diagnostic Trouble Codes NONE	Open/Closed Loop OPEN	Fuel Pump Relay (FP) OFF
Fuel Level Sensor 3.5 V	Fuel Tank Pressure Sensor 2.5 V	Transmission Fluid Temperature Sensor 3.5 V	Transmission Turbine Shaft Speed Sensor 0 mph
Transmission Range Switch PARK	Trans Pressure Control Solenoid 0	Transmission Shift Solenoid 1 ON	Transmission Shift Solenoid 2 OFF

Measured Ignition Timing °BTDC　　　Base Timing: 10°　　　Actual Timing: 20°

44.　The composite vehicle has entered the shop with a poor economy concern. Using the KOEO scan tool data shown in the above figure, determine the cause of the fuel economy concern.

　　A.　Faulty IAT sensor

　　B.　Faulty ECT sensor

　　C.　Faulty MAP sensor

　　D.　Faulty MAF sensor

45.　The composite vehicle will not start, and there is no signal from the crank sensor. Technician A says that the voltage to the sensor should be measured. Technician B says that the resistance of the sensor should be measured. Who is correct?

　　A.　A only

　　B.　B only

　　C.　Both A and B

　　D.　Neither A nor B

46.　The composite vehicle will not start, and the scan tool will not communicate. Technician A says that an unplugged instrument panel cluster will cause the scan tool to not communicate. Technician B says that a data bus that is shorted to ground will cause this problem. Who is correct?

　　A.　A only

　　B.　B only

　　C.　Both A and B

　　D.　Neither A nor B

47. The composite vehicle has a high idle. The cause of this condition could be:

    A. A sticking throttle cable

    B. ECM terminal 172 shorted to power

    C. ECM terminal 172 shorted to ground

    D. No voltage to the A/C pressure switch

48. The composite vehicle has been towed into the shop with an immobilizer system failure. This failure will:

    A. Not allow engine cranking

    B. Turn the security indicator on constantly

    C. Cause the engine to start and stall

    D. Not allow scan tool communication

49. The composite vehicle has entered the shop with a DTC P0440 – EVAP system leak. Before injecting smoke to look for the leak, the technician should:

    A. Command the purge solenoid closed

    B. Block off the EVAP canister

    C. Install a non-vented fuel cap

    D. Command the vent solenoid closed

50. The catalytic converter readiness monitor for the composite vehicle displays "Not Complete." This reading can be caused by:

    A. A failed catalytic converter monitor

    B. A failed oxygen sensor monitor

    C. Miscommunication of the scan tool

    D. A faulty ECM

PREPARATION EXAM 4

1. Technician A recommends testing the battery before reprogramming the electronic control module (ECM). Technician B recommends disabling the daytime running lamps while reprogramming. Who is correct?

    A. A only

    B. B only

    C. Both A and B

    D. Neither A nor B

2. Technician A says that installing a high-flow exhaust system will affect the operation of the exhaust gas recirculation (EGR) system. Technician B says an exhaust system of this type may not be legal. Who is correct?

   A.　A only

   B.　B only

   C.　Both A and B

   D.　Neither A nor B

3. After sitting overnight, a vehicle starts, stalls, and dies unless the throttle is pressed. After warming up, the vehicle will idle, but RPM is still low. What could cause this problem?

   A.　A dirty idle air control (IAC) valve

   B.　An intake manifold vacuum leak

   C.　A dirty air filter

   D.　A leaking fuel pump check valve

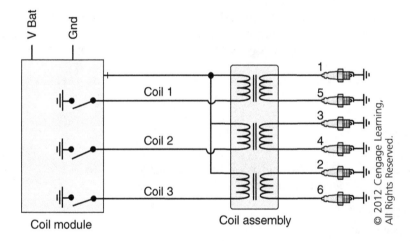

4. Technician A says that a fault in the ground circuit of one of the pictured coils will cause a miss on one cylinder. Technician B says that an open in the power wire will cause the vehicle not to start. Who is correct?

   A.　A only

   B.　B only

   C.　Both A and B

   D.　Neither A nor B

5. A vehicle that failed the I/M test for excessive NO$_x$ emissions will likely have to be tested under what vehicle condition to duplicate the concern?

    A. Idle

    B. Heavy acceleration

    C. Light acceleration

    D. Deceleration

6. A vehicle enters the shop with a DTC P0440 – EVAP system small leak detected. Technician A says that shop air can be used to pressurize the system as long as it is regulated down to 0.5 psi (3.4 kPa). Technician B says that spraying soapy water around a pressurized system will allow the technician to locate the leak. Who is correct?

    A. A only

    B. B only

    C. Both A and B

    D. Neither A nor B

7. A faulty IAT sensor is suspected of causing a poor fuel economy complaint. Technician A says that the sensor voltage should steadily increase as the sensor heats up. Technician B says that sensor resistance should steadily increase as the sensor heats up. Who is correct?

    A. A only

    B. B only

    C. Both A and B

    D. Neither A nor B

8. A poor-quality air filter has been installed in a vehicle, and the MAF sensor is now contaminated. Which code will this condition most likely set?

    A. P0174 – system rich

    B. P0171 – system lean

    C. P0101 – MAF sensor performance

    D. P0131 – O$_2$ sensor low voltage

9. Technician A says that a vacuum leak will cause lower-than-normal MAF sensor voltage at idle. Technician B says that a vacuum leak will cause lower-than-normal MAP sensor voltage at idle. Who is correct?

    A. A only

    B. B only

    C. Both A and B

    D. Neither A nor B

10. A long-term fuel trim of 20 percent is indicated on the scan tool. Technician A says that this indicates a rich condition. Technician B says that this could be the result of excessive fuel pressure caused by an unapproved aftermarket fuel pressure regulator. Who is correct?

    A. A only
    B. B only
    C. Both A and B
    D. Neither A nor B

11. A cracked spark plug will be displayed on the secondary voltage pattern as:

    A. A higher firing line
    B. A lower firing line
    C. A shorter dwell period
    D. A longer dwell period

12. To test the operation of an O_2 sensor, a vacuum line is removed. The voltage of the sensor should:

    A. Decrease
    B. Increase
    C. Fluctuate rapidly
    D. Show no change

13. A low voltage condition displayed by the O_2 sensor will LEAST LIKELY be caused by:

    A. Low fuel pressure
    B. A vacuum leak
    C. The EVAP purge valve stuck open
    D. A cylinder misfire

14. The results of a cracked exhaust manifold would be:

    A. Higher engine RPM
    B. Higher O_2 sensor voltage
    C. Lower O_2 sensor voltage
    D. Lower engine RPM

15. Technician A says that when CO levels are high the technician will have to diagnose a lean condition. Technician B says that when O_2 levels are high the technician will have to diagnose a rich condition. Who is correct?

    A. A only
    B. B only
    C. Both A and B
    D. Neither A nor B

16. The ignition rotor of a distributor ignition (DI) system has failed several times and created a no-start condition. The root cause of this failure could be which of the following?

    A. High plug wire resistance

    B. A cracked spark plug

    C. High coil wire resistance

    D. A cracked distributor cap

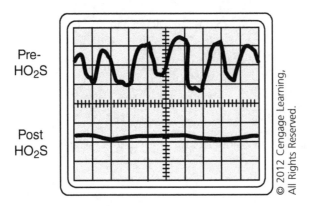

17. What do the pictured O_2 sensor patterns indicate?

    A. A malfunctioning post HO_2S

    B. A malfunctioning pre-HO_2S

    C. A functioning catalytic converter

    D. A malfunctioning catalytic converter

18. Technician A says that if all of the vehicle monitors have run and the MIL is not illuminated, all of the monitors have passed. Technician B says that if the MIL is illuminated, one or more of the monitors have failed. Who is correct?

    A. A only

    B. B only

    C. Both A and B

    D. Neither A nor B

19. A vehicle that continually goes into open loop at idle could be the result of:

    A. A continuously lean condition

    B. A continuously rich condition

    C. A faulty O_2 sensor heater

    D. Poor fuel quality

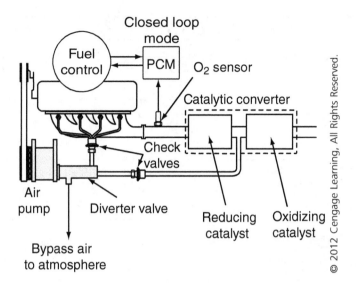

20. Under the conditions shown in the above image, where will the air output from the air pump go?

    A. Downstream

    B. Upstream

    C. To the air cleaner

    D. No air will be pumped

21. A vehicle starts and dies. Technician A says that a faulty IAC valve can cause this condition. Technician B says that a failure in the immobilizer system can cause this condition. Who is correct?

    A. A only

    B. B only

    C. Both A and B

    D. Neither A nor B

22. After installing a new intake manifold gasket, which of the following methods is the safest and most effective way to locate an intake manifold vacuum leak?

    A. Spray the intake with throttle body cleaner

    B. Spray the intake with a soapy water solution

    C. Blow compressed shop air into the intake

    D. Connect a smoke machine to a vacuum line

SCAN TOOL DATA

Engine Coolant Temperature (ECT) Sensor 0.51 volts	Intake Air Temperature (IAT) Sensor 2.4 volts	Manifold Absolute Pressure (MAP) Sensor 1.8 volts	Throttle Position Sensor (TPS) 40%
Engine Speed Sensor (rpm) 1,700 rpm	Air Fuel Ratio Sensor AFRS 1/1 - 3.7 volts AFRS 2/1 - 3.7 volts	Vehicle Speed Sensor (VSS) 55 mph	Battery Voltage (B+) 14.3 volts
Fuel Pressure (FP) - 1.4 volts	Evaporative Emission Canister Solenoid (EVAP) ON	Torque Converter Clutch (TCC) Solenoid ON	EGR Valve Control (EGR) 30 percent
Malfunction Indicator Lamp (MIL) ON	Diagnostic Trouble Codes P0171*, P0174**, P0087***	Open/Closed Loop CLOSED	Fuel Pump 100%

*P0171 - Bank 1 System Lean
**P0174 - Bank 2 System Lean
***P0087 - Fuel Pressusre (FP) Sensor Low

23. The composite vehicle has low power under most driving conditions. Using the scan tool data shown, which of these would cause this condition?

    A. A faulty A/F ratio sensor

    B. A faulty fuel injector

    C. A weak fuel pump

    D. A faulty Fuel Pressure (FP) sensor

24. A failed catalytic converter would likely cause which of the following exhaust emissions to be higher at idle?

    A. HC and CO

    B. HC and NO_x

    C. CO and CO_2

    D. CO_2 and O_2

25. What will the ECM do when the O_2 sensor shows an overall lean reading?

    A. Decrease fuel trim

    B. Increase idle speed

    C. Decrease fuel delivery

    D. Increase fuel trim

26. An engine coolant temperature sensor is being tested for operation. While the resistance is being tested as the engine is warming up, the value should:

    A. Steadily increase

    B. Steadily decrease

    C. Fluctuate rapidly

    D. Read O.L.

27. A vehicle has a rough idle and excessive HC emissions at idle, but it runs smoothly and has acceptable HC emissions at 2000 rpm. Which of the following could cause this condition?

   A. High spark plug wire resistance

   B. A clogged fuel filter

   C. A leaking intake runner gasket

   D. A fouled spark plug

28. The operation of the purge valve is being tested. With the purge valve open KOER, O_2 sensor voltage should:

   A. Increase

   B. Decrease

   C. Remain the same

   D. Fluctuate rapidly

29. A vehicle is brought to the shop for an illuminated MIL and has a stored DTC P0420 – catalyst efficiency. Technician A says to check the engine for oil consumption. Technician B says to replace the downstream O_2 sensor. Who is correct?

   A. A only

   B. B only

   C. Both A and B

   D. Neither A nor B

30. A contaminated MAF sensor will increase which of the following emissions?

   A. HC

   B. CO_2

   C. CO

   D. O_2

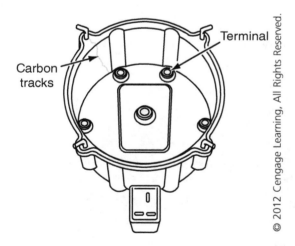

31. The carbon track in the distributor cap pictured above could be caused by:

   A. A cracked spark plug

   B. High coil wire resistance

   C. High plug wire resistance

   D. Insufficient plug gap

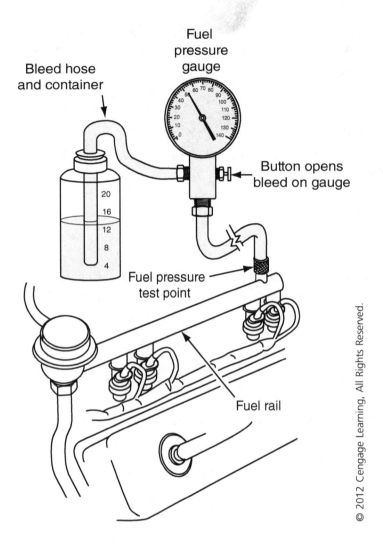

Bleed hose and container

Fuel pressure gauge

Button opens bleed on gauge

Fuel pressure test point

Fuel rail

32. What test is being performed in the above image?

    A. Fuel pump pressure
    B. Fuel pump volume
    C. Fuel injector balance
    D. Fuel injector flow

33. The TPS voltage is displayed as 0.5 V on the scan tool but measures 1.0 V at the sensor using a DVOM. The cause of this could be:

    A. High TPS signal wire resistance
    B. Low TPS reference voltage
    C. High TPS signal return resistance
    D. Low TPS sensor resistance

34. During a test of an electric air injection reaction (AIR) system, the pump is actuated with the scan tool and air is injected upstream. What does a low O_2 sensor voltage indicate under these conditions?

    A. A properly-operating system

    B. A malfunctioning diverter valve

    C. A malfunctioning AIR pump

    D. A malfunctioning O_2 sensor

35. Excessive no-load CO emissions would LEAST LIKELY be caused by:

    A. A clogged PCV valve

    B. A leaking fuel injector

    C. Excessive fuel pressure

    D. A fouled spark plug

36. All of the following should be done while preparing to reprogram an ECM EXCEPT:

    A. Turn on the radio

    B. Test the battery and generator

    C. Disable daytime running lamps

    D. Ensure scan tool communication

37. Excessive loaded HC emissions can be caused by which of the following?

    A. A clogged fuel filter

    B. A clogged air filter

    C. High plug wire resistance

    D. High fuel pressure

38. All of the following are symptoms of a sluggish O_2 sensor EXCEPT:

    A. Reduced fuel economy

    B. Poor idle quality

    C. Hesitation on acceleration

    D. Lower exhaust emissions

39. The composite vehicle has entered the shop with the MIL lamp illuminated and DTCs P0014 and P0021 – intake valve timing control fault banks 1 and 2. Technician A says that this could be caused by an open cam position solenoid. Technician B says that this could be caused by low oil pump output. Who is correct?

    A. A only

    B. B only

    C. Both A and B

    D. Neither A nor B

40. A stuck open PCV valve could cause all of the following EXCEPT:

     A. Increased engine idle speed

     B. Decreased fuel trim readings

     C. Rough idle concerns

     D. Multiple cylinder misfire

41. The throttle system is being diagnosed on the composite vehicle. Technician A says that two TP sensors that display identical voltages are used in the event that one fails. Technician B says that two accelerator pedal position (APP) sensors that display identical voltages are used in the event that one fails. Who is correct?

     A. A only

     B. B only

     C. Both A and B

     D. Neither A nor B

42. A vacuum leak at the injector o-ring could cause any of these EXCEPT:

     A. Black exhaust smoke

     B. Low O_2 sensor voltage

     C. Rough unstable idle

     D. Spark knock

43. The composite vehicle has DTC P0301 – cylinder #1 misfire. The ignition coil for cylinder one has no spark. An oscilloscope is used to measure the voltage at terminal 14 of the ECM and displays steady system voltage. What component has failed?

     A. The ignition coil

     B. The spark plug

     C. The ECM

     D. The plug wire

44. The composite vehicle is brought into the shop with a low power concern and a fuel pressure measurement of 38 psi (262 kPa). Technician A says that directing the return line into a container while actuating the fuel pump will test the pump output. Technician B says that the fuel filter should be checked for a restriction. Who is correct?

     A. A only

     B. B only

     C. Both A and B

     D. Neither A nor B

45. The composite vehicle's MAP sensor is reading 2.5 V at idle. This could be caused by:

     A. A restricted air filter

     B. A carbon-clogged MAP sensor

     C. An open MAP sensor signal wire

     D. A restricted exhaust

46. The composite vehicle enters the shop with a DTC P0452 – fuel tank pressure sensor low voltage. Which of the following could cause this condition?

   A. An open at ECM terminal 104
   B. An open at ECM terminal 110
   C. A short to ground at ECM terminal 104
   D. A short to ground at ECM terminal 110

47. The composite vehicle will not start and the fuel pressure (FP) sensor reads 0.5 V. There are no DTCs present. This could be caused by:

   A. A blown fuse #22
   B. An open at terminal 660 of the FPCM
   C. A faulty fuel pump
   D. A faulty fuel pressure (FP) sensor

48. The composite vehicle will not start. Technician A says that the resistance of the crank sensor cannot be measured and that waveform diagnosis should be used. Technician B says that the crank sensor will produce a square wave signal. Who is correct?

   A. A only
   B. B only
   C. Both A and B
   D. Neither A nor B

49. The composite vehicle has failed an I/M idle test for excessive CO levels. Technician A says that this could be caused by a leaking fuel pressure regulator. Technician B says that this could be caused by putting too much fuel in the fuel tank. Who is correct?

   A. A only
   B. B only
   C. Both A and B
   D. Neither A nor B

50. The EGR valve position sensor on the composite vehicle is being tested KOEO. With the valve in the fully open position, the sensor should display what voltage reading?

   A. 0.5 V
   B. 1.0 V
   C. 4.0 V
   D. 4.5 V

PREPARATION EXAM 5

1. A throttle position sensor (TPS) is fixed at 4.8 V. Technician A says that the electronic control module (ECM) will ignore this reading and illuminate the MIL. Technician B says that the ECM will rely on other engine sensors to perform engine control operations. Who is correct?

   A. A only
   B. B only
   C. Both A and B
   D. Neither A nor B

2. The engine coolant temperature (ECT) sensor reads erratic after the engine has warmed up. Technician A says that there may be air in the cooling system. Technician B says that ignition cables may be routed near the signal wire, giving a false reading. Who is correct?

   A. A only

   B. B only

   C. Both A and B

   D. Neither A nor B

SCAN TOOL DATA

Engine Coolant Temperature Sensor (ECT)	Intake Air Temperature Sensor (IAT)	MAP Sensor 1.8 V	Throttle Position Sensor (TPS)
0.46 Volts	2.24 Volts	MAF Sensor 1.1 V	0.6 Volts
Engine Speed Sensor (RPM) 1500 rpm	Heated O$_2$ HO$_2$S Upstream 0.2–0.5V Downstream 0.1–0.3V	Vehicle Speed Sensor (VSS) 0 mph	Battery Voltage (B+) 14.2 Volts
Idle Air Control Valve (IAC) 0 percent	Evaporative Emission Canister Solenoid (EVAP) OFF	Torque Converter Clutch Solenoid (TCC) OFF	EGR Valve Control Solenoid (EGR) 0 percent
Malfunction Indicator Lamp (MIL) OFF	Diagnostic Trouble Codes NONE	Open/Closed Loop CLOSED	Fuel Pump Relay (FP) On
Fuel Level Sensor 3.5 V	Fuel Tank Pressure Sensor 2.0 V	Transmission Fluid Temperature Sensor 0.6 V	Transmission Turbine Shaft Speed Sensor 0 mph
Transmission Range Switch PARK	Trans Pressure Control Solenoid 80%	Transmission Shift Solenoid 1 ON	Transmission Shift Solenoid 2 OFF

Measured Ignition Timing °BTDC   Base Timing: 10°   Actual Timing: 20°

3. What diagnosis can be made using the scan tool data in the above figure?

   A. A leaking MAP sensor vacuum hose

   B. A faulty IAC valve

   C. An intake manifold vacuum leak

   D. An exhaust manifold leak

4. How will the ECM react to high knock sensor voltage?

   A. Increase the coil's dwell

   B. Advance ignition timing

   C. Decrease the coil's dwell

   D. Retard ignition timing

5. Which of the following tests would be the most beneficial if a vehicle had failed the I/M test for excessive CO emissions?

   A. Power balance test

   B. Fuel pressure test

   C. Exhaust backpressure

   D. Compression test

6. Which of the following symptoms could LEAST LIKELY be associated with an EGR valve stuck open?

   A. Hard starting

   B. Rough idle

   C. Engine stalling

   D. Detonation

7. When the ECM receives a low voltage signal from the HO_2S sensor it will:

   A. Increase fuel delivery

   B. Decrease fuel delivery

   C. Increase engine RPM

   D. Decrease engine RPM

8. A faulty catalytic converter has been diagnosed. Technician A says that a malfunctioning PCV system could have caused the converter to fail. Technician B says that leaking valve stem seals could have caused the converter to fail. Who is correct?

   A. A only

   B. B only

   C. Both A and B

   D. Neither A nor B

9. A customer is concerned with a "rotten egg" smell coming from the exhaust, with no other drivability concerns. What should the technician do to correct this problem?

   A. Replace the catalytic converter

   B. Replace the O_2 sensors

   C. Suggest an alternate fuel source

   D. Suggest a complete tune-up

10. A vehicle is brought into the shop with a DTC P0455 – EVAP system large leak detected. Technician A says to first check the gas cap to see if it is tight. Technician B says that a lack of vacuum to the purge valve can cause a leak detection. Who is correct?

    A. A only

    B. B only

    C. Both A and B

    D. Neither A nor B

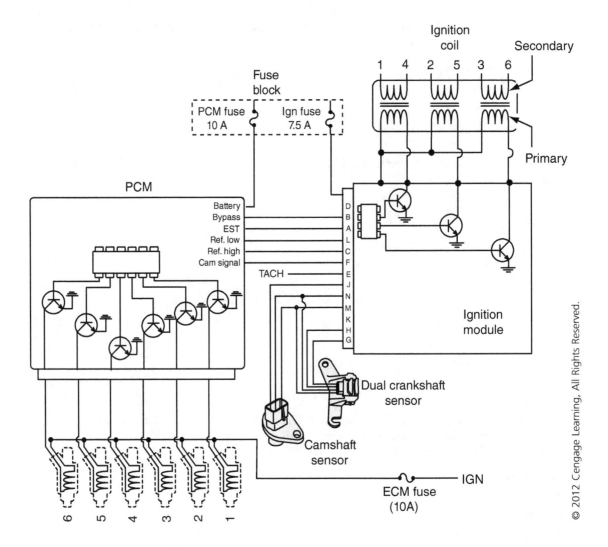

11. The vehicle using the engine control system above will not start. The vehicle has spark but no injector pulse. Technician A says that the camshaft position (CMP) sensor could be faulty. Technician B says that the crankshaft position (CKP) sensor could be faulty. Who is correct?

    A. A only

    B. B only

    C. Both A and B

    D. Neither A nor B

12. The oxygen sensor heater is being tested. Technician A says that measuring the voltage drop of the heater will verify its operation. Technician B says that when it is cold, a functioning heater will cause O_2 sensor voltage to drop below 100 mV when it reaches operating temperature without the engine running. Who is correct?

    A. A only

    B. B only

    C. Both A and B

    D. Neither A nor B

13. A failure of which emission system would cause increased NO_x emissions?

    A.   Evaporative emissions (EVAP)

    B.   Positive crankcase ventilation (PCV)

    C.   Early fuel evaporation (EFE)

    D.   Exhaust gas recirculation (EGR)

14. A rich O_2 sensor voltage would be:

    A.   Below 450 mV

    B.   Above 450 mV

    C.   Between 250 mV and 750 mV

    D.   Fixed at 500 mV

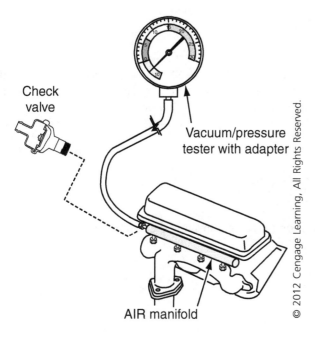

Check valve

Vacuum/pressure tester with adapter

AIR manifold

15. What is being tested in the above picture?

    A.   Exhaust backpressure

    B.   Check valve function

    C.   AIR pump pressure

    D.   EGR system flow

16. A distributor ignition (DI) vehicle is brought into the shop with a DTC P0340 – CMP sensor/ CKP sensor correlation. Technician A says that this can be caused by a worn distributor gear. Technician B says that this could be caused by worn cam timing components. Who is correct?

    A.   A only

    B.   B only

    C.   Both A and B

    D.   Neither A nor B

17. The O_2 sensor voltage on a vehicle is fixed at 0.03 V. When the exhaust gas is measured directly with an exhaust gas analyzer, what gas will be high if the O_2 sensor voltage is correct?

    A. CO

    B. NO_x

    C. CO_2

    D. O_2

18. A vehicle is hard to start on cold mornings after sitting all night. Technician A says that the vehicle will have to sit overnight before diagnosis can begin. Technician B says that the fuel may pressure should be monitored for any pressure loss when the vehicle is pulled into the service bay. Who is correct?

    A. A only

    B. B only

    C. Both A and B

    D. Neither A nor B

19. A vehicle fails an emission test for excessive NO_x at 2500 rpm. Technician A says that this could be caused by clogged EGR passages. Technician B says that this could be caused by a diverter valve stuck in the downstream position. Who is correct?

    A. A only

    B. B only

    C. Both A and B

    D. Neither A nor B

20. The hoses of an air injection reaction (AIR) system show signs of burning. This could be a result of:

    A. A faulty air pump

    B. A faulty diverter valve

    C. A faulty check valve

    D. A restricted exhaust

21. Dielectric grease is used on many connectors and terminals when a vehicle's electrical system is being serviced. Technician A says that the grease will help keep water out of the connectors. Technician B says that the grease will allow for a better connection because it is conductive. Who is correct?

    A. A only

    B. B only

    C. Both A and B

    D. Neither A nor B

22. A burned exhaust valve is likely to increase which two of the following emissions?

   A. CO_2 and CO

   B. HC and O_2

   C. NO_x and CO

   D. HC and CO

SCAN TOOL DATA			
Engine Coolant Temperature Sensor (ECT) 0 Volts	Intake Air Temperature Sensor (IAT) 2.9 Volts	MAP Sensor 4.5 V MAF Sensor 0.2 V	Throttle Position Sensor (TPS) 0.6 Volts
Engine Speed Sensor (RPM) 0 rpm	Heated O_2 HO_2S Upstream 0 V Downstream 0 V	Vehicle Speed Sensor (VSS) 0 mph	Battery Voltage (B+) 12.6 Volts
Idle Air Control Valve (IAC) 30 percent	Evaporative Emission Canister Solenoid (EVAP) OFF	Torque Converter Clutch Solenoid (TCC) OFF	EGR Valve Control Solenoid (EGR) 0 percent
Malfunction Indicator Lamp (MIL) ON	Diagnostic Trouble Codes P0117	Open/Closed Loop OPEN	Fuel Pump Relay (FP) OFF
Fuel Level Sensor 3.3 V	Fuel Tank Pressure Sensor 4.5 V	Transmission Fluid Temperature Sensor 1.8 V	Transmission Turbine Shaft Speed Sensor 0 mph
Transmission Range Switch PARK	Trans Pressure Control Solenoid 0	Transmission Shift Solenoid 1 OFF	Transmission Shift Solenoid 2 OFF
Measured Ignition Timing °BTDC	Base Timing: 10°	Actual Timing: 0	

23. Based on the KOEO scan tool data shown in the above figure, the DTC P0117 would LEAST LIKELY be caused by which of the following?

   A. ECT sensor shorted

   B. ECT sensor signal circuit shorted to ground

   C. A faulty ECM

   D. ECT sensor open

24. Which of the following gases would be elevated if a rich condition were present?

   A. CO_2

   B. NO_x

   C. CO

   D. O_2

25. A restricted catalytic converter is suspected on a vehicle, but the O_2 sensors are too difficult to reach in order to measure backpressure. Which of the following tools can be used to verify the restricted converter?

    A. An oscilloscope

    B. A digital multi-meter (DMM)

    C. A compression gauge

    D. A vacuum gauge

26. A vehicle cuts out and stumbles at times and has a DTC, P0122 – TPS low voltage stored in its history. Technician A says that the sensor voltage should be viewed with a scope. Technician B says that a sudden voltage drop to zero displayed on the scope indicates that the sensor is shorted. Who is correct?

    A. A only

    B. B only

    C. Both A and B

    D. Neither A nor B

27. A catalytic converter is being tested after replacement, and the inlet and outlet temperatures are measured. What should the temperatures show?

    A. A hotter inlet temperature

    B. A hotter outlet temperature

    C. The same temperatures should be measured

    D. Measuring the temperature is irrelevant

28. A common method of tampering involves installing a resistor in series with the intake air temperature (IAT) sensor. What affect will this have on the sensor voltage?

    A. IAT voltage will increase

    B. IAT voltage will decrease

    C. Sensor voltage will be bypassed

    D. Sensor voltage will remain steady and high

29. The scan tool is displaying a long-term fuel trim of (LTFT) −25 percent. What is the ECM trying to compensate for?

    A. A lean condition

    B. A rich condition

    C. An advanced distributor

    D. A retarded distributor

30. Which of the following would prevent the EVAP monitor from running?

    A. A leaking fuel cap

    B. Faulty purge valve

    C. Low fuel level

    D. A vacuum leak

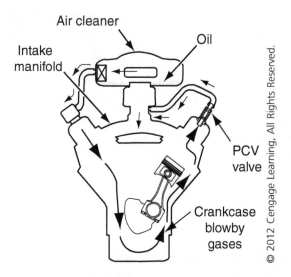

31. Oil is found in the air filter of the engine pictured above. Technician A says that the PCV valve may not be functional. Technician B says that the rings may be worn. Who is correct?

    A. A only

    B. B only

    C. Both A and B

    D. Neither A nor B

32. Refer to the following idle speed emissions levels to determine the possible fault:

    HC – 40 ppm

    CO – 4.3 percent

    CO_2 – 10 percent

    O_2 – 0.1 percent

    A. A restricted fuel filter

    B. An intake vacuum leak

    C. Insufficient fuel pressure

    D. Stuck open EVAP purge valve

33. A stuck closed PCV valve could cause which of the following conditions?

    A. Excessive blowby

    B. Oil in the intake pipe

    C. High idle speed

    D. Lower exhaust emissions

34. A stuck open thermostat will increase which of the following emissions?

    A. CO

    B. CO_2

    C. NO_x

    D. O_2

35. All of the following can cause the O_2 sensor to become contaminated with oil EXCEPT:

    A. Leaking valve guides
    B. PCV system faults
    C. Worn piston rings
    D. Leaking rear main seal

36. A symptom of an air injection system directing air upstream at all times would be:

    A. Random cylinder misfire
    B. Increased fuel consumption
    C. Low power output
    D. Engine backfire on acceleration

37. A vehicle failed the no-load I/M test for excessive HC emissions; however, it passed the loaded test for HCs. Which of the following could cause this condition?

    A. A vacuum leak
    B. A fouled spark plug
    C. Clogged EGR passages
    D. A clogged fuel filter

38. A vehicle with distributorless electronic ignition (EI) has an engine miss under heavy load but runs okay at idle. All of the following could be the cause of the miss EXCEPT:

    A. A carbon tracked spark plug
    B. Excessive plug wire resistance
    C. A faulty ignition module
    D. A faulty ignition coil

39. The composite vehicle enters the shop with a DTC P0031 – Bank 1 AFR sensor heater circuit malfunction. Which of the following would be the best first step in diagnosis?

    A. Replace the AFR sensor
    B. Check for a blown fuse 20
    C. Check the heater resistance
    D. Check for a damaged ECM terminal 152

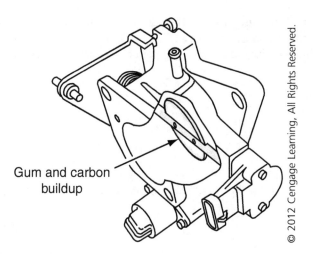

Gum and carbon buildup

40. The problem indicated in the figure above can cause all of the following problems EXCEPT:

   A. High idle

   B. Sticking throttle

   C. Stalling

   D. Rough idle

41. The composite vehicle is slow to refuel. Technician A says that a clogged purge solenoid will cause this condition. Technician B says that a stuck open vent solenoid can cause this condition. Who is correct?

   A. A only

   B. B only

   C. Both A and B

   D. Neither A nor B

42. The composite vehicle has poor acceleration and an illuminated MIL. Technician A says that one failed TPS can cause this. Technician B says that an open at terminal "a" of the TPS will cause this condition. Who is correct?

   A. A only

   B. B only

   C. Both A and B

   D. Neither A nor B

43. The composite vehicle has an illuminated MIL and DTCs P0011 and P0021 – intake valve timing control fault banks 1 and 2. This could be caused by all of the following EXCEPT:

   A. Low oil level

   B. Incorrect oil viscosity

   C. A faulty cam sensor

   D. Low oil pump output

44. Technician A says that if an invalid key is used to start the composite vehicle, the ECM will disable the fuel injectors within two seconds. Technician B says that if an invalid key is used to start the composite vehicle, the security indicator will remain on constantly. Who is correct?

    A. A only
    B. B only
    C. Both A and B
    D. Neither A nor B

45. Which of the following sensors, if faulty, will prevent the composite vehicle from starting?

    A. CMP sensor
    B. CKP sensor
    C. Knock sensor (KS)
    D. MAP sensor

46. When testing the EVAP system on the composite vehicle, the fuel tank pressure sensor remains at 1 V and slowly increases after the vent and purge solenoids have been commanded off. This could indicate:

    A. A stuck closed purge solenoid
    B. A stuck open vent solenoid
    C. A clogged purge hose
    D. A clogged vent hose

47. After a repair, the keep alive memory (KAM) in the ECM can be reset by all of the following EXCEPT:

    A. Clearing ECM DTCs
    B. Removing the ECM fuse
    C. Unplugging the ECM
    D. Disconnecting the battery

48. The composite vehicle will not start, so the fuel pump enable circuit is being tested. While measuring the resistance from ECM terminal 127 to fuel pump control module (FPCM) terminal 649, the ohmmeter displays O.L. This reading indicates:

    A. A good fuel pump enable circuit
    B. A bad fuel pump control module (FPCM)
    C. An open fuel pump
    D. A faulty fuel pump relay

49. One ignition coil on the composite vehicle is suspected of causing a misfire. Which of the following would be an acceptable primary coil resistance reading?

    A.   0.1 Ω

    B.   1 Ω

    C.   10 KΩ

    D.   O.L.

50. Which sensor on the composite vehicle is used primarily to detect exhaust flow in the EGR system?

    A.   EGR position

    B.   Throttle position

    C.   MAP sensor

    D.   MAF sensor

PREPARATION EXAM 6

1. One injector driver has failed several times. What should be checked to find the cause of this repeated failure?

    A.   Injector relay voltage drop

    B.   Injector resistance

    C.   ECM battery voltage

    D.   Injector ground resistance

2. A rich air/fuel ratio will cause HO_2S voltage to be:

    A.   Below 100 mV

    B.   Above 100 mV

    C.   Above 450 mV

    D.   Fixed at 450 mV

3. Which of the following customer concerns might be present with a stuck closed exhaust gas recirculation (EGR) valve?

    A.   Detonation

    B.   Engine stalling

    C.   Rough idle

    D.   Low power output

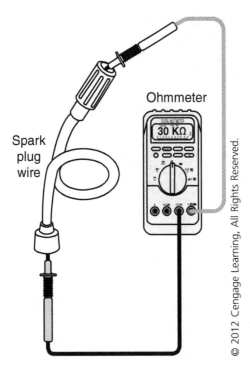

Ohmmeter

30 KΩ

Spark
plug
wire

4. The meter shown above displays 30 KΩ. Technician A says that the wire has too low
    resistance. Technician B says that the displayed reading would cause the firing line to be
    lower than normal when viewed on a scope. Who is correct?

    A. A only
    B. B only
    C. Both A and B
    D. Neither A nor B

5. A catalytic converter has failed several times with efficiency problems. Technician A says that
    this could be caused by an exhaust manifold leak. Technician B says that this could be caused
    by a seeping head gasket. Who is correct?

    A. A only
    B. B only
    C. Both A and B
    D. Neither A nor B

6. O_2 sensor voltage is fixed high. Which of the following emissions gases would be expected to
    increase?

    A. O_2
    B. CO
    C. NO_x
    D. CO_2

7. Technician A says that a vehicle will be rejected from the I/M test if one or more of the monitors have not run. Technician B says that a vehicle will be rejected if the MIL does not work. Who is correct?

   A. A only

   B. B only

   C. Both A and B

   D. Neither A nor B

8. A vehicle has entered the shop with a DTC P0401 – insufficient EGR flow. Which of the following modifications can cause this condition?

   A. Installing a high-flow exhaust system

   B. Installing a high-flow air intake system

   C. Installing a larger throttle body

   D. Installing a larger MAF sensor

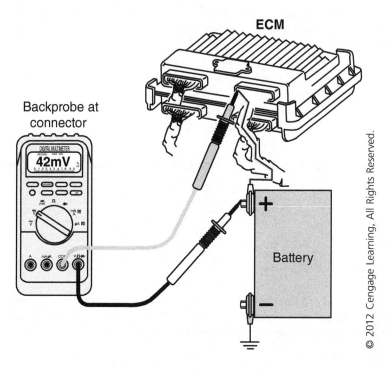

9. The ECM power circuit is being tested as shown above. Technician A says that the circuit has high resistance since the meter is not displaying 12 V. Technician B says that corrosion in the switch can cause a reading that is higher than 0 V. Who is correct?

   A. A only

   B. B only

   C. Both A and B

   D. Neither A nor B

10. The AIR injection tube has rusted off of the vehicle's catalytic converter. Which two emissions will be increased by this fault?

    A. HC and CO_2
    B. CO and NO_x
    C. HC and CO
    D. O_2 and CO_2

11. A vehicle equipped with a distributor ignition (DI) exhibits spark knock. Which of the following conditions could cause this?

    A. Too-advanced ignition timing
    B. Too-retarded ignition timing
    C. Switched plug wires
    D. High resistance in the plug wires

12. A vehicle has been brought to the shop with a concern of lowered fuel economy. Technician A says that the use of oxygenated fuels such as RFG (reformulated gasoline) can cause lowered fuel economy. Technician B says that the use of fuels with high alcohol content can cause lowered fuel economy. Who is correct?

    A. A only
    B. B only
    C. Both A and B
    D. Neither A nor B

13. A vehicle starts and immediately dies. Which of the following is the most likely cause of this condition?

    A. TPS voltage above 4.0 V
    B. An immobilizer fault
    C. Low fuel pressure
    D. Contaminated MAF sensor

14. A vehicle has failed a loaded I/M test for excessive HC emissions and a misfire. Technician A says that a restricted fuel filter could cause this condition. Technician B says that a fault in the secondary ignition system could cause this condition. Who is correct?

    A. A only
    B. B only
    C. Both A and B
    D. Neither A nor B

15. A vehicle has failed the I/M emissions test for excessive NO_x output. Which of the following conditions is most likely to cause elevated NO_x emissions?

    A. Retarded ignition timing
    B. A high-flow air intake system
    C. Debris-plugged radiator
    D. Excessive fuel pressure

16. An engine with a DI has no spark and will not start. A test light is connected to the negative coil circuit and flashes while the vehicle is being cranked. Which of the following components could be faulty?

    A. Crank sensor

    B. Ignition module

    C. Ignition fuse

    D. Ignition coil

17. The HO_2S heater circuit is being tested with an ohmmeter. What would a reading of O.L. indicate?

    A. The circuit is open

    B. The circuit is shorted

    C. There is excessive corrosion

    D. The circuit is functional

18. Which of the following would cause an increase in CO emissions?

    A. An intake manifold gasket leak

    B. Clogged injectors

    C. A contaminated MAF sensor

    D. Excessive fuel pressure

19. A vehicle has entered the shop several times with a plugged fuel filter and now has entered the shop with a faulty fuel pump. Technician A says that the tank should be cleaned before assembly. Technician B says that the customer should find a different fuel supplier. Who is correct?

    A. A only

    B. B only

    C. Both A and B

    D. Neither A nor B

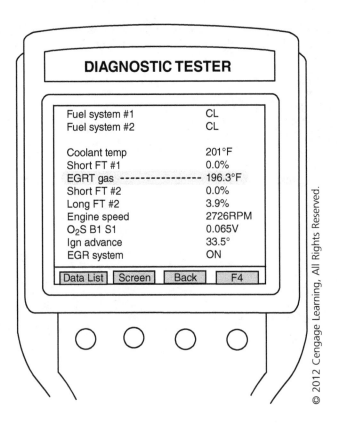

20. While the EGR valve is being commanded open, the scan tool data in the above figure is displayed. Technician A says that the EGR temperature sensor is faulty. Technician B says that this condition will illuminate the MIL. Who is correct?

    A. A only

    B. B only

    C. Both A and B

    D. Neither A nor B

21. The ECT sensor is being tested for operation. Technician A says that sensor resistance should increase as its temperature increases. Technician B says that the sensor voltage should increase as its temperature increases. Who is correct?

    A. A only

    B. B only

    C. Both A and B

    D. Neither A nor B

22. What effect can low-quality air filters have on MAF equipped vehicles?

    A. Cause a rich condition

    B. Increase NO_x emissions

    C. Cause a low idle speed

    D. Cause a rough idle

23. During routine maintenance, a technician discovers that the air filter is soaked with oil. Which of the following would be the most likely cause of this problem?

    A. A plugged PCV valve

    B. An overfilled crankcase

    C. A plugged PCV vent

    D. Excessive oil pressure

24. The voltage drop of the ECM ground circuit is 0.8 V. What does this indicate?

    A. A short in the circuit

    B. An open in the circuit

    C. High resistance in the circuit

    D. Low resistance in the circuit

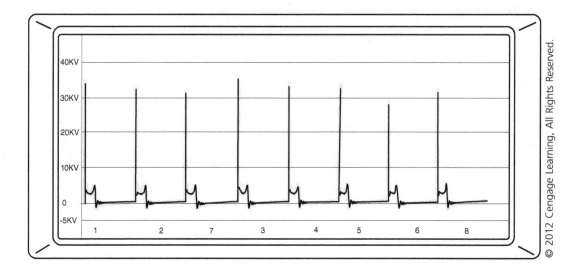

25. Which of the following could cause the secondary ignition waveforms shown above?

    A. Faulty coil wire

    B. Faulty plug wire

    C. Fouled spark plug

    D. Rich air/fuel mixture

26. An SFI vehicle will not start. Neither spark nor injector pulse is present. Technician A says that a TPS voltage fixed above 4.0 V will cause this condition. Technician B says that a faulty CKP sensor can cause this condition. Who is correct?

    A. A only

    B. B only

    C. Both A and B

    D. Neither A nor B

27. Technician A says that high CO emissions are a good indicator of a lean condition.
    Technician B says that high O_2 levels are a good indicator of a rich condition. Who is correct?

    A. A only

    B. B only

    C. Both A and B

    D. Neither A nor B

28. An exhaust manifold leak can cause which of the following:

    A. High O_2 sensor readings

    B. Negative fuel trim numbers

    C. Low O_2 sensor readings

    D. Low MAF sensor readings

29. A vehicle has a rough and unstable idle. A check of the exhaust shows elevated HCs, CO_2 at
    11 percent, O_2 at six percent, and CO less than one percent. Which of the following is most
    likely to cause these conditions?

    A. A vacuum leak

    B. A leaking injector

    C. A dirty throttle body

    D. A faulty fuel pump check valve

30. The fuel injectors are being current tested on a vehicle, and one is found to be high. Which of
    the following can cause this?

    A. High resistance

    B. Low resistance

    C. Low voltage

    D. High fuel pressure

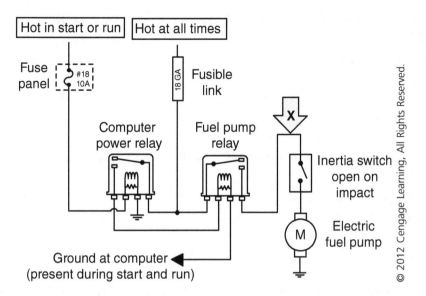

31. A vehicle is brought to the shop for a no-start concern, and it has been determined that the fuel pump is not running. In the above figure, the voltage at point X in the circuit, measured while the fuel pump is commanded on, is 0 V. Which of the following could cause this problem?

    A. A stuck closed fuel pump relay

    B. An open fuel pump relay coil

    C. An open fuel pump ground

    D. A tripped inertia switch

32. Technician A says that during open loop, the vehicle's fuel needs are calculated using engine sensors such as the MAF, TPS, and ECT. Technician B says that during closed loop, fuel calculations are made based on O_2 sensor readings. Who is correct?

    A. A only

    B. B only

    C. Both A and B

    D. Neither A nor B

33. Negative fuel trim values can be caused by all of the following EXCEPT:

    A. A contaminated MAF sensor

    B. Excessive fuel pressure

    C. A leaking fuel injector

    D. A leaking EVAP purge valve

34. Which of the following would LEAST LIKELY cause a vehicle to immediately fail the I/M test?

    A. No MIL illumination
    B. An illuminated MIL
    C. A missing fuel cap
    D. Low fuel level

35. A vehicle has failed the I/M test for excessive CO at idle. Which of the following problems could cause this condition?

    A. A manifold vacuum leak
    B. A restricted fuel filter
    C. A restricted catalytic converter
    D. A faulty AIR pump

36. A catalytic converter is suspected faulty. Which of the following emissions would be LEAST LIKELY to increase?

    A. HC
    B. NO_x
    C. CO
    D. CO_2

37. The composite vehicle has entered the shop with a DTC P0011 – camshaft over advanced bank 1. Technician A says that this could be caused by debris stuck in the camshaft position actuator control solenoid. Technician B says that this could be caused by low oil pressure. Who is correct?

    A. A only
    B. B only
    C. Both A and B
    D. Neither A nor B

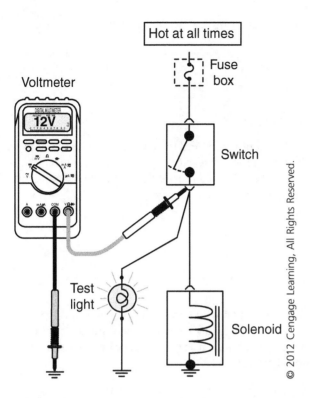

38. The solenoid in the circuit is inoperative. While testing with the meter shown in the above figure, the circuit displays 12 V, and the test light illuminates when the switch is closed. All of the following can cause this EXCEPT:

    A. An open ground circuit

    B. An open solenoid

    C. An open wire between the switch and the solenoid

    D. An open switch

39. A late-model vehicle with on-board refueling vapor recovery (ORVR) is slow to refuel. Which of the following could cause this problem?

    A. A restricted vent filter

    B. A stuck open vent solenoid

    C. A stuck closed purge solenoid

    D. A restricted purge line

40. A vehicle failed the I/M test for excessive CO emissions and its O_2 sensor readings are fixed above 850 mV. All of the following could cause this EXCEPT:

    A. ECT sensor

    B. Skewed O_2 sensor

    C. Leaky fuel pressure regulator

    D. A skewed MAP sensor

41. A catalytic converter can be damaged by all of the following EXCEPT:

    A. A faulty O_2 sensor

    B. A blown head gasket

    C. A faulty plug wire

    D. A dirty air filter

42. Intermittently, the composite vehicle will not start, and the anti-theft indicator lamp will flash. Technician A says that the battery in the key may be getting low. Technician B says that key rings containing other transponder keys for different vehicles can affect the signal received by the antenna. Who is correct?

    A. A only

    B. B only

    C. Both A and B

    D. Neither A nor B

43. The composite vehicle is brought into the shop with a miss on cylinder 1 and a suspected faulty ignition coil. Technician A says that the resistance from coil terminal "A" to terminal "B" should be 10 KΩ. Technician B says that the resistance from terminal "A" to terminal "C" should be 1 Ω. Who is correct?

    A. A only

    B. B only

    C. Both A and B

    D. Neither A nor B

44. The brake pedal position switch of the composite vehicle is being tested. The voltage drop across the switch with the pedal depressed is 12.6 V. This voltage indicates:

    A. Low switch input voltage

    B. A closed brake switch

    C. An open brake switch

    D. High power circuit resistance

45. The composite vehicle will enter closed loop when:

    A. Valid AFR sensor readings are received by the ECM

    B. Coolant temperature reaches 195°F (91°C)

    C. The engine has run for at least ten minutes

    D. TPS is less than 50 percent

46. The composite vehicle will not start, and the CKP sensor is suspected faulty. To check the sensor's output while cranking, what setting should the DVOM be set to?

    A. AC amps

    B. DC amps

    C. AC volts

    D. DC volts

47. Which of the following would be an acceptable resistance measurement of the composite vehicle's fuel injector?

    A. 5 Ω

    B. 10 Ω

    C. 10 KΩ

    D. O.L.

48. The composite vehicle has entered the shop with a DTC P0440 – EVAP system leak. Technician A says that the system can be leak tested with a smoke machine connected to shop air. Technician B says that the vent solenoid must be commanded shut. Who is correct?

    A. A only

    B. B only

    C. Both A and B

    D. Neither A nor B

49. The composite vehicle has failed the I/M test for excessive NO_x emissions. Which of the following could cause this condition?

    A. An open at ECM terminal 108

    B. An open at ECM terminal 110

    C. An open at ECM terminal 160

    D. An open at ECM terminal 240

50. The composite vehicle's EVAP system is being tested. When the engine is running and the purge and vent solenoids are commanded on, the fuel tank pressure sensor voltage steadily decreases. This drop in voltage indicates:

    A. Insufficient vacuum supply

    B. A stuck open vent valve

    C. A faulty fuel tank pressure sensor

    D. A sealed EVAP system

Answer Keys and Explanations

INTRODUCTION

Included in this section are the answer keys for each preparation exam, followed by individual, detailed answer explanations and a reference identifying the designated task area being assessed by each specific question. This additional reference information may prove useful if you need to refer back to the task list located in Section 4 of this book for additional support.

PREPARATION EXAM 1—ANSWER KEY

1.	A	21.	A	41.	C
2.	C	22.	C	42.	B
3.	C	23.	B	43.	D
4.	D	24.	C	44.	C
5.	A	25.	D	45.	B
6.	B	26.	A	46.	D
7.	C	27.	C	47.	B
8.	A	28.	D	48.	A
9.	D	29.	B	49.	A
10.	B	30.	A	50.	B
11.	D	31.	D		
12.	B	32.	B		
13.	A	33.	A		
14.	B	34.	B		
15.	D	35.	B		
16.	C	36.	C		
17.	D	37.	A		
18.	A	38.	D		
19.	C	39.	D		
20.	C	40.	B		

PREPARATION EXAM 1—EXPLANATIONS

1. Technician A says that switching oil viscosities may affect the operation of the variable valve timing (VVT) system. Technician B says that switching oil viscosities may affect the operation of the positive crankcase ventilation (PCV) system. Who is correct?

TASK A.7

   A. A only
   B. B only
   C. Both A and B
   D. Neither A nor B

   Answer A is correct. Only Technician A is correct. Since most VVT systems operate using oil pressure, using an oil of the wrong viscosity may not allow the camshaft timing mechanisms to operate properly.

   Answer B is incorrect. Technician B is incorrect. Using an oil of the wrong viscosity will not affect the operation of the PCV system.

   Answer C is incorrect. Only Technician A is correct.

   Answer D is incorrect. Technician A is correct.

2. An engine misfire cannot be duplicated. Technician A says that spraying a saltwater solution on the spark plug wires may cause the misfire to occur easier. Technician B says that the customer should be quizzed specifically about when the misfire occurs. Who is correct?

TASK C.1

   A. A only
   B. B only
   C. Both A and B
   D. Neither A nor B

   Answer A is incorrect. Technician B is also correct.

   Answer B is incorrect. Technician A is also correct.

   Answer C is correct. Both Technicians are correct. Spraying the ignition system with a saltwater solution can often make it easier to duplicate an ignition-related misfire. It is also important for the technician to gather as much information from the customer about the concern when trying to duplicate intermittent problems.

   Answer D is incorrect. Technician A is correct.

3. Technician A says that maintaining the proper battery voltage during electronic control module (ECM) reprogramming is critical. Technician B says that if battery voltage falls too low during reprogramming the ECM may have to be replaced. Who is correct?

TASK B.6

   A. A only
   B. B only
   C. Both A and B
   D. Neither A nor B

   Answer A is incorrect. Technician B is also correct.

   Answer B is incorrect. Technician A is also correct.

   Answer C is correct. Both Technicians are correct. When reprogramming the ECM, battery voltage must remain above a certain point. If battery voltage falls below that point during the programming process, the ECM may need to be replaced on some vehicles.

   Answer D is incorrect. Both Technicians are correct.

TASK D.5

4. A vehicle equipped with a 4-cylinder engine has come into the shop with a diagnostic trouble code (DTC) P0172 – bank 1 rich. Which of the following would LEAST LIKELY cause this problem?

    A.   Excessive fuel pressure

    B.   A plugged fuel return line

    C.   A leaking injector

    D.   A plugged fuel filter

    Answer A is incorrect. Excessive fuel pressure would cause a rich condition.

    Answer B is incorrect. A plugged fuel return line would cause a rich condition and cause the DTC P0172 to set.

    Answer C is incorrect. A leaking fuel injector could cause a rich condition to occur.

    Answer D is correct. A plugged fuel filter would cause a lean condition and would not set the DTC P0172.

TASK B.5

5. A vehicle is brought into the shop with a DTC P0171 – System Lean, and a vacuum leak is suspected. Technician A says that a smoke machine can be used to locate the vacuum leak. Technician B says that the smoke should be blown around the intake with the engine running while noting an increase in RPM. Who is correct?

    A.   A only

    B.   B only

    C.   Both A and B

    D.   Neither A nor B

    Answer A is correct. Only Technician A is correct. The smoke machine can be used to inject smoke into the vehicle's intake. The smoke should be injected with the engine off. The leak can be found by identifying where the smoke escapes.

    Answer B is incorrect. The smoke will not cause the engine RPM to raise if injected into a running engine.

    Answer C is incorrect. Only Technician A is correct.

    Answer D is incorrect. Technician A is correct.

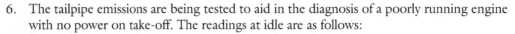

TASK F.7

6. The tailpipe emissions are being tested to aid in the diagnosis of a poorly running engine with no power on take-off. The readings at idle are as follows:

    HC – 120 ppm

    CO – 0.5 percent

    CO_2 – 12 percent

    O_2 – 10 percent

    What could be causing these readings?

    A.   A leaking injector

    B.   Low fuel pressure

    C.   A restricted air inlet

    D.   Stuck open exhaust gas recirculation (EGR) valve

    Answer A is incorrect. A leaking injector would cause a rich condition, which would be identified by high CO levels in the exhaust.

    Answer B is correct. Low fuel pressure would cause a lean condition, which can be identified by the high oxygen levels in the exhaust.

    Answer C is incorrect. A restricted air inlet may cause a slightly rich mixture, which would increase CO levels.

    Answer D is incorrect. EGR gas would take the place of oxygen entering the cylinder if it were stuck open. In this event, oxygen gases would be low.

7. The air injection reaction (AIR) system hoses have burned repeatedly on a vehicle. What component should the technician check to prevent this failure again?

TASK E.15

   A. AIR pump

   B. AIR diverter valve

   C. Exhaust check valves

   D. Catalytic converter

Answer A is incorrect. A faulty AIR pump could cause excessive CO and HC emissions, but it would not cause the hoses to burn.

Answer B is incorrect. A faulty diverter valve could cause higher emissions but not burned hoses.

Answer C is correct. Burned air injection hoses are a sign of a faulty exhaust check valve. Replacing only the hoses will result in a comeback.

Answer D is incorrect. Even if a catalytic converter were restricted, the check valves should still be able to keep exhaust pressure from entering the AIR system hoses.

8. Technician A says that low manifold vacuum can be seen on the scan tool as high manifold absolute power (MAP) sensor voltage readings. Technician B says a clogged exhaust can cause low MAP sensor voltage readings. Who is correct?

TASK B.9

   A. A only

   B. B only

   C. Both A and B

   D. Neither A nor B

Answer A is correct. Only Technician A is correct. Low intake manifold vacuum will cause MAP sensor readings to be higher than normal.

Answer B is incorrect. A clogged exhaust would cause lower-than-normal intake manifold vacuum, which would cause higher-than-normal MAP sensor readings.

Answer C is incorrect. Only Technician A is correct.

Answer D is incorrect. Technician A is correct.

9. A restricted catalytic converter has been diagnosed. This could be caused by:

TASK A.12

   A. A malfunctioning EGR system

   B. A clogged injector

   C. Excessive combustion chamber deposits

   D. Leaking valve guide seals

Answer A is incorrect. A malfunctioning EGR system will not cause damage to the catalytic converter.

Answer B is incorrect. A clogged injector would cause a misfire; however, there would not be excessive HCs in the exhaust to cause damage to the catalytic converter.

Answer C is incorrect. Excessive combustion chamber deposits would cause higher-than-normal compression ratios but would not cause the catalytic converter to deteriorate.

Answer D is correct. Leaking valve guide seals, especially on the exhaust valves, will allow oil to get to the catalytic converter, which could cause it to overheat and melt down or to become coated with oil ash deposits, restricting its flow.

SCAN TOOL DATA

Engine Coolant Temperature Sensor (ECT) 0.46 Volts	Intake Air Temperature Sensor (IAT) 2.24 Volts	MAP Sensor 1.8 V / MAF Sensor 1.1 V	Throttle Position Sensor (TPS) 0.6 Volts
Engine Speed Sensor (RPM) 1500 rpm	Heated O₂ HO₂S Upstream 0.2–0.5 V Downstream 0.1–0.3 V	Vehicle Speed Sensor (VSS) 0 mph	Battery Voltage (B+) 14.2 Volts
Idle Air Control Valve (IAC) 0 percent	Evaporative Emission Canister Solenoid (EVAP) OFF	Torque Converter Clutch Solenoid (TCC) OFF	EGR Valve Control Solenoid (EGR) 0 percent
Malfunction Indicator Lamp (MIL) OFF	Diagnostic Trouble Codes NONE	Open/Closed Loop CLOSED	Fuel Pump Relay (FP) On
Fuel Level Sensor 3.5 V	Fuel Tank Pressure Sensor 2.0 V	Transmission Fluid Temperature Sensor 0.6 V	Transmission Turbine Shaft Speed Sensor 0 mph
Transmission Range Switch PARK	Trans Pressure Control Solenoid 80%	Transmission Shift Solenoid 1 ON	Transmission Shift Solenoid 2 OFF

Measured Ignition Timing °BTDC   Base Timing: 10°   Actual Timing: 20°

TASK B.8

10. The scan tool data in the figure above is being reviewed. Technician A says that the data could indicate low fuel pressure. Technician B says that the data could indicate an intake manifold vacuum leak. Who is correct?

A. A only

B. B only

C. Both A and B

D. Neither A nor B

Answer A is incorrect. The low O_2 sensor voltage readings indicate a lean condition, which could be caused by low fuel pressure. However, the MAP sensor voltage is slightly higher than normal, indicating high intake manifold pressure, which could be caused by an intake manifold vacuum leak.

Answer B is correct. Only Technician B is correct. The high RPM, closed throttle position sensor (TPS) voltage, and 0 percent IAC are indicators of an intake manifold vacuum leak.

Answer C is incorrect. Only Technician B is correct.

Answer D is incorrect. Technician B is correct.

11. Technician A says that high O_2 emissions levels are an indicator of an efficient running engine. Technician B says that high CO_2 levels are an indicator of a rich condition. Who is correct?

TASK F.5

    A. A only

    B. B only

    C. Both A and B

    D. Neither A nor B

    Answer A is incorrect. High O_2 levels are an indicator of a lean-running engine.

    Answer B is incorrect. High CO_2 levels are an indicator of an efficiently running engine.

    Answer C is incorrect. Neither Technician is correct.

    Answer D is correct. Neither Technician is correct.

12. An oil type air filter can cause contamination of the MAF sensor, which will:

TASK D.2

    A. Cause a rich condition

    B. Cause a lean condition

    C. Cause EGR system faults

    D. Cause high CO_2 emissions

    Answer A is incorrect. A contaminated MAF sensor will cause a lean condition.

    Answer B is correct. If the MAF sensor is contaminated, unmetered air will enter the engine. The ECM will only deliver fuel for the amount of air that is measured. This will cause the air-to-fuel ratio to increase and be lean.

    Answer C is incorrect. An oil type air filter will not have any effect on the EGR system.

    Answer D is incorrect. High CO_2 readings occur when the engine is operating at a 14.7:1 air-to-fuel ratio. Any other air/fuel ratio will cause CO_2 levels to lower.

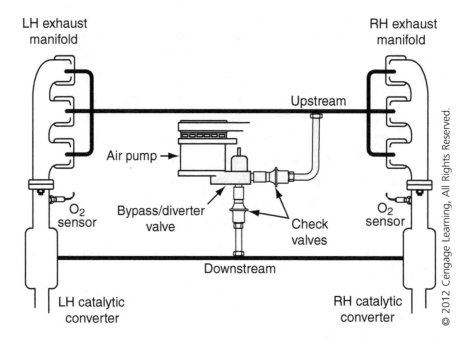

LH exhaust manifold

RH exhaust manifold

Upstream

Air pump →

Bypass/diverter valve

O_2 sensor

Check valves

O_2 sensor

Downstream

LH catalytic converter

RH catalytic converter

TASK E.8

13. The diverter valve in the AIR system pictured in the above figure has failed and continuously pumps air upstream. This will cause:

  A. Low O_2 sensor voltages

  B. High O_2 sensor voltages

  C. Catalyst damage

  D. A rough idle

  Answer A is correct. The excess oxygen being injected upstream would cause the upstream oxygen sensors to produce a low voltage signal.

  Answer B is incorrect. High oxygen content in the exhaust causes the O_2 sensors to produce a low voltage signal.

  Answer C is incorrect. The excess oxygen in the exhaust will not cause damage to the converter.

  Answer D is incorrect. The AIR system's injecting air into the exhaust will have no affect on the engine's idle quality.

TASK C.5

14. The output of a two wire crank sensor can be tested most accurately by using:

  A. An ohmmeter

  B. An AC voltmeter

  C. A DC voltmeter

  D. A test light

  Answer A is incorrect. While an ohmmeter is a useful tool when checking the resistance of a two wire crank sensor, it will not give any information about the output signal of the sensor.

  Answer B is correct. A two wire crank sensor generates an AC signal, which can be tested with an AC voltmeter.

  Answer C is incorrect. A two wire crank sensor will generate an AC voltage signal, which will not show the sensor's accurate voltage output.

  Answer D is incorrect. A test light is not the appropriate tool for testing a crank sensor or any other sensor for that matter.

15. Technician A says that a plugged injector can be diagnosed by performing a fuel pressure test. Technician B says that the plugged injector could have been caused by debris in the tank. Who is correct?

TASK D.12

    A. A only

    B. B only

    C. Both A and B

    D. Neither A nor B

Answer A is incorrect. A plugged injector cannot be diagnosed by performing a fuel pressure test. An injector balance test that actually measures pressure drop as the injector is being pulsed is the proper test for diagnosing a clogged injector.

Answer B is incorrect. Fuel filters will trap any debris that is large enough to clog an injector.

Answer C is incorrect. Neither Technician is correct.

Answer D is correct. Neither Technician is correct. Any debris that clogs an injector may have come from fuel line corrosion due to excessive alcohol content in the fuel. Excessive buildup of mineral deposits on the injector tip could also cause this.

SCAN TOOL DATA

Engine Coolant Temperature (ECT) Sensor 0.52 volts	Intake Air Temperature (IAT) Sensor 2.4 volts	Manifold Absolute Pressure (MAP) Sensor 1.9 volts	Throttle Position Sensor (TPS) 0.5 volts
Engine Speed Sensor (rpm) 450 rpm	Heated Oxygen Sensor (HO₂S) 0.3–0.8 volts	Vehicle Speed Sensor (VSS) 0 mph	Battery Voltage (B+) 12.8 volts
Idle Air Control (IAC) Valve 100 percent	Evaporative Emission Canister Solenoid (EVAP) OFF	Torque Converter Clutch (TCC) Solenoid OFF	EGR Valve Control Solenoid (EGR) 0 percent
Malfunction Indicator Lamp (MIL) OFF	Diagnostic Trouble Codes NONE	Open/Closed Loop CLOSED	Fuel Pump Relay (FP) ON

Measured Ignition Timing °BTDC   Base Timing: 10°   Actual Timing: 11°

16. A technician is attempting to verify the concern of dying at idle. Based on the above scan tool data, what could be wrong with the vehicle?

TASK B.1

    A. Faulty TPS

    B. A charging system fault

    C. Faulty IAC

    D. Faulty MAP sensor

Answer A is incorrect. The TPS is reading the correct closed throttle voltage.

Answer B is incorrect. Although charging system voltage is low, it is due to low engine RPM.

Answer C is correct. The IAC is being commanded 100 percent, and the idle speed is only 450 rpm. The MAP sensor also indicates that vacuum is lower than normal due to the low RPM.

Answer D is incorrect. The MAP sensor voltage is high due to the lack of air being supplied to the engine.

TASK E.5

17. An evaporative emissions (EVAP) system leak is being diagnosed, and the system is being smoke tested. Technician A says that the EVAP purge valve will have to be commanded shut since it is normally an open valve. Technician B says that the EVAP vent valve is normally closed and will therefore not have to be commanded shut. Who is correct?

   A. A only
   B. B only
   C. Both A and B
   D. Neither A nor B

   Answer A is incorrect. The EVAP purge valve is normally a closed valve and will not have to be commanded shut when leak testing the system.

   Answer B is incorrect. The EVAP vent solenoid is normally open and will have to be commanded shut when leak testing.

   Answer C is incorrect. Neither Technician is correct. Only the EVAP vent solenoid will have to be commanded shut when leak testing the EVAP system.

   Answer D is correct. Neither Technician is correct.

TASK C.7

18. A vehicle is brought into the shop with an illuminated MIL, a P0305 – cylinder 5 misfire, and a P0302 – cylinder 2 misfire on a V6 engine. The engine uses a waste spark ignition system, and the firing order is 1-2-3-4-5-6. The engine runs smooth with no evidence of a misfire; however, while monitoring the misfire counter, cylinders 2 and 5 show a continuous misfire under all running conditions. This could be caused by:

   A. A damaged crankshaft tone wheel
   B. High resistance in the plug wires
   C. A faulty ECM coil driver
   D. A faulty 2/5 ignition coil

   Answer A is correct. A damaged tooth on the crankshaft tone wheel would show up as a change in engine speed, which would be interpreted as a misfire by the ECM. On a V-type engine, the misfire would be shown for two companion cylinders.

   Answer B is incorrect. Since the engine shows no physical signs of an engine miss, the plug wires should be considered good.

   Answer C is incorrect. If the ECM coil driver were faulty, there would be a noticeable engine miss.

   Answer D is incorrect. Since the engine shows no physical signs of an engine miss, the coil should be considered good.

TASK A.9

19. An engine with a vacuum leak will have what effect on exhaust emissions?

   A. Lower-than-normal HCs
   B. Lower-than-normal O_2
   C. Higher-than-normal O_2
   D. Higher-than-normal CO_2

   Answer A is incorrect. A vacuum leak will not lower HC emissions. If the leak were severe enough, a lean misfire would cause increased HC emissions.

   Answer B is incorrect. A vacuum leak will cause higher-than-normal O_2 levels.

   Answer C is correct. A vacuum leak will cause an excess of O_2 in the combustion chamber. This leftover O_2 will exit the engine through the exhaust.

   Answer D is incorrect. When the engine is operating at its peak efficiency, CO_2 levels are high. Any emissions-affecting problem will lower CO_2 levels.

20. A vehicle is brought into the shop with an illuminated MIL. The ECM has stored a DTC P0135 – HO_2S heater performance. The voltage drop of the power circuit is 12.1 V. This voltage could indicate:

TASK B.14

 A. Low system voltage

 B. An open O_2 sensor heater circuit

 C. Resistance in the power circuit

 D. An open O_2 heater ground circuit

Answer A is incorrect. Even if there were low system voltage, the voltage drop of the power circuit would still be very low.

Answer B is incorrect. Since the voltage drop is on the power circuit, this indicates that the problem is with the power circuit, not with the sensor.

Answer C is correct. A voltage drop of 12.1 V indicates that there is high resistance in the power circuit of the O_2 sensor heater circuit. The voltage drop of this circuit should be very close to zero.

Answer D is incorrect. If the O_2 sensor heater were open, a voltage drop of the power circuit would measure zero volts.

21. Technician A says that the MIL lamp should illuminate for approximately two seconds after the key is turned on. Technician B says that the MIL lamp should illuminate as soon any engine-related DTC is set. Who is correct?

TASK F.13

 A. A only

 B. B only

 C. Both A and B

 D. Neither A nor B

Answer A is correct. Only Technician A is correct. The MIL should illuminate during the bulb check for approximately two seconds when the key is turned on.

Answer B is incorrect. Some engine-related DTCs will not turn on the MIL. Other DTCs must fail twice before the MIL will illuminate.

Answer C is incorrect. Only Technician A is correct.

Answer D is incorrect. Technician A is correct.

22. A vehicle fails a loaded I/M test for excessive HC emissions. This could be caused by:

TASK F.10

 A. Leaking vacuum hose

 B. Fouled spark plug

 C. Cracked spark plug

 D. Hung open EVAP purge valve

Answer A is incorrect. A leaking vacuum hose would be more likely to cause high HC emissions at idle rather than under a load.

Answer B is incorrect. A fouled spark plug would cause an engine miss under all operating conditions, not just under a load.

Answer C is correct. A cracked spark plug would typically cause a miss when the engine is loaded and cause excessive HC emissions.

Answer D is incorrect. A hung open EVAP purge valve would cause high HC emissions at idle as well as under a load.

TASK A.6

23. A slipped timing belt is suspected on a poorly running, non-interference engine. What vehicle data would support this diagnosis?

    A.   Low O_2 sensor voltage

    B.   High MAP sensor voltage

    C.   A faulty cam sensor signal

    D.   High MAF sensor signal

Answer A is incorrect. A slipped timing belt would likely cause the O_2 sensors to read rich, which would be high voltage.

Answer B is correct. Retarded cam timing would cause the engine to have lower-than-normal vacuum. This would be shown as higher-than-normal MAP sensor voltage.

Answer C is incorrect. Even though the cam timing would be retarded, the sensor signal would still be present.

Answer D is incorrect. Since the engine would not be breathing as well due to the retarded cam, airflow into the engine would be less and would cause a lower-than-normal MAF sensor signal.

TASK B.13

24. A vehicle starts, runs for two seconds and dies. The LEAST LIKELY cause of this condition could be:

    A.   An immobilizer system fault

    B.   A faulty IAC valve

    C.   Low fuel pressure

    D.   A dirty throttle body

Answer A is incorrect. Most vehicles that have an immobilizer system fault will exhibit the given symptoms.

Answer B is incorrect. A faulty IAC valve will often cause symptoms similar to those described.

Answer C is correct. A vehicle is likely to idle even if fuel pressure is low. If the pressure is excessively low, the vehicle is not likely to start at all.

Answer D is incorrect. Like the IAC valve, if the throttle body were dirty, the vehicle may exhibit starting problems and low RPM idling concerns.

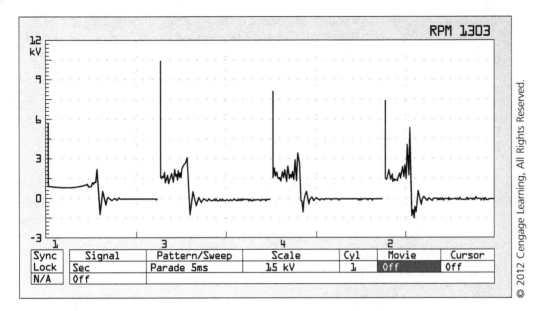

25. The secondary voltage patterns shown in the above figure were taken after performing a tune-up. The waveform could be caused by:

    A. A rich condition

    B. A bridged #3 plug gap

    C. Excessive plug wire resistance

    D. A too-small #1 plug gap

    TASK C.15

    Answer A is incorrect. A rich condition would cause all firing lines to be slightly lower than normal, not just one.

    Answer B is incorrect. While a bridged plug gap would cause lower firing voltages, the low voltage is on cylinder #1.

    Answer C is incorrect. Excessive plug wire resistance would cause high firing voltages.

    Answer D is correct. A plug that is gapped too small will have a low firing line and a longer spark line when it is scoped.

26. An ECM has been replaced several times for a non-functioning injector and a DTC P0201 – injector #1 control circuit. The injector will function for a while with the new ECM, but will soon fail again. Which test could find the cause of this failure?

    A. Injector resistance

    B. Injector voltage drop

    C. Control circuit resistance

    D. Power circuit voltage drop

    TASK B.22

    Answer A is correct. When an ECM is replaced for a driver malfunction, the actuator that is controlled by that driver should be checked for resistance. If the actuator is shorted, the excess current can damage the driver.

    Answer B is incorrect. Even if the injector is shorted, as long as it is the only load in the circuit, it will still drop 12 volts.

    Answer C is incorrect. Either high resistance or a short in the injector control circuit will not cause the injector driver to fail. High resistance would lower the current, which will not damage the driver, and a short to ground would cause the injector to stay on at all times.

    Answer D is incorrect. A large voltage drop in the power circuit would cause the circuit to have lower-than-normal current, which would not damage the injector driver in the ECM.

TASK E.10

27. Technician A says that a poorly maintained cooling system can cause excessive NO_x formation. Technician B says that plugged EGR passages can also cause excessive NO_x formation. Who is correct?

A. A only

B. B only

C. Both A and B

D. Neither A nor B

Answer A is incorrect. Technician B is also correct.

Answer B is incorrect. Technician A is also correct.

Answer C is correct. Both Technicians are correct. A poorly maintained cooling system can lead to higher engine operating temperatures, which will lead to excessive NO_x formation. Since the EGR system controls the formation of NO_x, a malfunction in that system could lead to excessive NO_x formation.

Answer D is incorrect. Both Technicians are correct.

TASK D.11

28. Fuel pressure is being monitored on a fuel pressure gauge. Technician A says that the pressure should go down when the vacuum line is removed from the fuel pressure regulator. Technician B says that fuel should drain from the regulator when the vacuum line is removed. Who is correct?

A. A only

B. B only

C. Both A and B

D. Neither A nor B

Answer A is incorrect. Fuel pressure should not decrease when the vacuum line is disconnected from the fuel pressure regulator.

Answer B is incorrect. There should not be any fuel leaking from the fuel pressure regulator when the vacuum hose is removed. Any leaking fuel indicates that the diaphragm has ruptured and the regulator needs to be replaced.

Answer C is incorrect. Neither Technician is correct.

Answer D is correct. Neither Technician is correct. Fuel pressure should increase when the vacuum line is disconnected, and no fuel should leak out of the vacuum hose or the fuel pressure regulator.

TASK E.6

29. Which piece of scan tool data would be the MOST helpful when trying to diagnose a plugged catalytic converter?

A. RPM

B. MAP

C. HO_2S

D. TPS

Answer A is incorrect. RPM would not be very useful when trying to diagnose a plugged catalytic converter.

Answer B is correct. If the catalytic converter were plugged, intake manifold vacuum would be low, which would cause higher MAP sensor readings.

Answer C is incorrect. A plugged catalytic converter would not have any noticeable effect on the HO_2S reading.

Answer D is incorrect. A plugged catalytic converter would have no effect on the TPS reading.

30. The O_2 sensor is reading low voltage (below 250 mV). What exhaust gas readings can be expected?

    A. High O_2 content

    B. High CO_2 content

    C. Low NO_x content

    D. Low O_2 content

    TASK B.19

    Answer A is correct. A reading of below 250 mV indicates that the system is running lean. High O_2 levels will accompany a lean-running engine.

    Answer B is incorrect. CO_2 content would be lower than expected since the engine is not operating at the optimum 14.7:1 air/fuel ratio.

    Answer C is incorrect. NO_x readings will increase slightly as the air/fuel ratio increases above 14.7:1.

    Answer D is incorrect. O_2 levels will rise as the air/fuel ratio becomes lean.

31. The current draw of a fuel pump is lower than specified. This could be caused by:

    A. Shorted fuel pump windings

    B. A stuck closed pressure regulator

    C. A clogged fuel filter

    D. Excessive circuit resistance

    TASK B.15

    Answer A is incorrect. Shorted fuel pump windings will increase the current flow of the circuit.

    Answer B is incorrect. A stuck closed fuel pressure regulator will cause the pump to work harder, which will increase the current flow of the circuit.

    Answer C is incorrect. A clogged fuel filter will cause the pump to work harder, which will increase the current flow of the circuit.

    Answer D is correct. Excessive resistance in the fuel pump circuit will lower the current flow in that circuit.

32. During a plug-in emissions test, the catalyst monitor displays "Not Complete." What is the possible cause?

    A. A failed catalytic converter

    B. The ECM memory has been reset

    C. The engine is not up to operating temperature

    D. An exhaust manifold leak

    TASK F.2

    Answer A is incorrect. A failed catalytic converter will still show a message of "Complete" even if it has failed.

    Answer B is correct. A message of "Not Complete" indicates that the ECM memory has been cleared and the catalyst monitor has not met the enabling criteria to run.

    Answer C is incorrect. As long as the catalyst monitor has run since the last time the ECM memory was cleared, the monitor will display as complete. Current engine operating conditions will not affect previously-run monitors.

    Answer D is incorrect. While an exhaust manifold leak may cause the monitor to fail, it will still display as complete.

TASK E.12

33. A malfunction of the PCV system is likely to cause all of these problems EXCEPT:

 A. Excessive blowby

 B. An illuminated MIL

 C. Rough/unstable idle speed

 D. Engine oil leaks

Answer A is correct. The PCV system will have no effect on how much blow-by the engine produces. Engine wear will determine how much blowby is produced.

Answer B is incorrect. A malfunction in the PCV system could cause the MIL to illuminate, especially if the vacuum hose is leaking.

Answer C is incorrect. A leaking vacuum hose on the PCV system can cause a rough or unstable idle.

Answer D is incorrect. A malfunctioning PCV system will allow pressure to build in the crankcase. This pressure pushes oil out of any weak or deteriorated gaskets, resulting in an oil leak.

TASK D.5

34. Positive fuel trim numbers can be caused by:

 A. High fuel pressure

 B. Clogged fuel filter

 C. Stuck closed fuel pressure regulator

 D. An injector hung open

Answer A is incorrect. High fuel pressure will cause a rich condition, which will cause fuel trim numbers to go negative.

Answer B is correct. A clogged fuel filter will cause a lean condition, which will cause fuel trim numbers to go positive.

Answer C is incorrect. A stuck closed fuel pressure regulator will cause fuel pressure to increase, resulting in a rich condition. This would cause fuel trim numbers to go negative.

Answer D is incorrect. An injector that is hung open will result in a rich condition, which will cause fuel trim numbers to go negative.

TASK B.17

35. While being tested by a scan tool, TPS voltage drops to near zero (0.02 V). This indicates:

 A. The throttle is at idle position

 B. A bad spot in the TPS

 C. The scan tool is processing data

 D. Normal activity

Answer A is incorrect. Even at idle, the TPS will normally read at least 0.5 V. Typically, a reading lower than that indicates a bad spot in the sensor.

Answer B is correct. If TPS voltage drops to near zero during its sweep, it is likely that the TPS has a bad spot.

Answer C is incorrect. Even though many scan tools process data at a much slower rate than the vehicle, the data displayed will be the last data sample taken.

Answer D is incorrect. In most cases, it should never be considered normal for the TPS voltage to drop below 0.3 V.

36. A vehicle has failed an emissions test with high NO_x levels, and tampering is suspected. The emission system that was most likely tampered with would be:

TASK E.2

    A. EVAP system

    B. PCV system

    C. EGR system

    D. ORVR system

    Answer A is incorrect. The EVAP system is designed to control HC emissions from the fuel tank.

    Answer B is incorrect. The PCV system is designed to control HC emissions from the crankcase.

    Answer C is correct. If the EGR system is tampered with, NO_x emissions may increase.

    Answer D is incorrect. The on-board refueling vapor recovery (ORVR) system is designed to control HC emissions from the fuel tank while refueling.

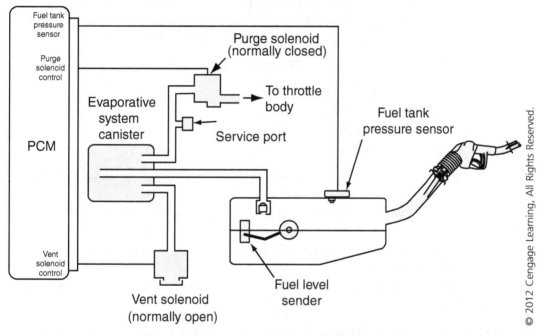

37. The customer has a concern of a slow fill when refueling the fuel system in the above figure, and the pump nozzle keeps kicking off. This could be caused by:

TASK E.7

    A. A stuck closed vent solenoid

    B. A leaking hose from the tank to the canister

    C. A leaking purge solenoid

    D. A stuck closed purge solenoid

    Answer A is correct. A stuck closed vent solenoid would not allow vapors inside the tank to escape as it was being filled. This would cause pressure to build in the tank and kick the fueling nozzle off.

    Answer B is incorrect. A leaking hose would allow vapors to escape the tank and would not slow the filling process.

    Answer C is incorrect. A leaking purge solenoid would allow vapors to escape the tank and would not slow the filling process.

    Answer D is incorrect. The purge solenoid is normally closed and will not affect filling performance.

TASK B.5

38. A vehicle is brought into the shop with a rough idle and both upstream O_2 sensors displaying high voltage. Which of the following could be the cause?

   A.   Faulty O_2 sensors
   B.   An intake manifold vacuum leak
   C.   Restricted air filter
   D.   Stuck open EGR valve

   Answer A is incorrect. It is unlikely that both O_2 sensors would go bad at the same time, but should this occur, the ECM would recognize the lack of switching and set a code, which would not cause a rough idle.

   Answer B is incorrect. While an intake manifold vacuum leak could cause a rough idle, it would be a lean condition, which would cause the O_2 sensors to read low voltage.

   Answer C is incorrect. While a restricted air filter could cause a rich condition on a speed density type fuel controlled system, it is unlikely to cause a rough idle concern.

   Answer D is correct. A stuck open EGR valve will cause a rough idle, and the lack of oxygen getting into the engine will cause the O_2 sensors to read rich.

TASK F.11

39. A vehicle fails a loaded emissions test for excessive CO. This could be caused by:

   A.   An intake manifold vacuum leak
   B.   Low fuel pump output
   C.   Clogged fuel injectors
   D.   Plugged fuel return line

   Answer A is incorrect. An intake manifold leak would cause a lean condition, which would not increase CO emissions.

   Answer B is incorrect. Low fuel pressure would cause a lean condition, which would not increase CO emissions.

   Answer C is incorrect. Clogged fuel injectors would cause a lean condition, which would not increase CO emissions.

   Answer D is correct. A plugged fuel return line would cause a rich condition, which would elevate CO readings.

TASK E.12

40. All of the following emission system faults could cause a rough idle EXCEPT:

   A.   A stuck open EGR valve
   B.   A stuck open EVAP vent valve
   C.   A stuck open EVAP purge valve
   D.   A leaking PCV vacuum hose

   Answer A is incorrect. A stuck open EGR valve would cause a rough idle concern.

   Answer B is correct. A stuck open vent valve would not cause any idle concerns.

   Answer C is incorrect. A stuck open EVAP purge valve would cause a rough idle concern.

   Answer D is incorrect. A leaking PCV vacuum hose would cause a rough idle concern, especially on a MAF controlled vehicle.

41. An improperly maintained cooling system is likely to cause what problem?

    A. Poor fuel economy

    B. Rough idle

    C. Detonation

    D. Low power output

    TASK A.8

    Answer A is incorrect. An improperly maintained cooling system is likely to cause higher engine temperatures. Typically lower engine temperatures lead to poor fuel economy.

    Answer B is incorrect. An improperly maintained cooling system is not likely to cause any idle concerns.

    Answer C is correct. An improperly maintained cooling system is likely to cause higher engine temperatures, which could lead to detonation.

    Answer D is incorrect. An improperly maintained cooling system will not likely have an effect on the vehicle's power output.

42. All of the following can cause excessive CO tailpipe emissions EXCEPT:

    A. A leaking injector

    B. A leaking vacuum hose

    C. Excessive fuel pressure

    D. A hanging EVAP purge solenoid

    TASK F.9

    Answer A is incorrect. A leaking injector would cause a rich condition that would increase CO emissions.

    Answer B is correct. A leaking vacuum hose would create a lean condition, which would increase oxygen levels in the exhaust, not CO.

    Answer C is incorrect. Excessive fuel pressure would cause a rich condition that would increase CO emissions.

    Answer D is incorrect. A hanging open EVAP purge solenoid would cause a rich condition that would increase CO emissions.

43. The EVAP system for the composite vehicle is being diagnosed. Technician A says that the EVAP system monitor will run no matter what the level of fuel in the tank. Technician B says that the EVAP system monitor will run no matter what the engine temperature. Who is correct?

    TASK B.7

    A. A only

    B. B only

    C. Both A and B

    D. Neither A nor B

    Answer A is incorrect. The EVAP system monitor for the composite vehicle will only run if the fuel level is between ¼ and ¾ full.

    Answer B is incorrect. The EVAP system monitor will only run if the engine temperature is below 86°F (30°C).

    Answer C is incorrect. Neither Technician is correct.

    Answer D is correct. Neither Technician is correct.

TASK A.3

44. The composite vehicle's VVT system:

   A. Changes the timing of the intake cam
   B. Changes the timing of the exhaust cam
   C. Changes the timing of both cams
   D. Changes the cam's duration

Answer A is incorrect. The VVT system of the composite vehicle changes the valve overlap by advancing or retarding the intake camshaft.

Answer B is incorrect. The VVT system does not change the timing of the exhaust cam.

Answer C is correct. The VVT system changes the timing of both the intake and exhaust camshaft.

Answer D is incorrect. The VVT system does not change the duration of the camshaft.

TASK F.12

45. The composite vehicle fails a loaded I/M test for excessive NO_x emissions. The possible cause for this condition could be:

   A. A ruptured EGR vacuum line
   B. A bad connection at ECM terminal 108
   C. A bad connection at ECM terminal 28
   D. A leaking EGR valve diaphragm

Answer A is incorrect. The composite vehicle does not use a vacuum controlled EGR valve.

Answer B is correct. ECM terminal 108 controls the EGR valve solenoid. If there were a bad connection at this terminal, the EGR valve would not function.

Answer C is incorrect. ECM terminal 28 is the EGR position sensor input from the EGR position sensor. If there were a bad connection at this point, the ECM would not know the position of the EGR valve.

Answer D is incorrect. The composite vehicle uses an electronic EGR valve.

TASK B.3

46. The composite vehicle is brought into the shop with a DTC P0172 – Fuel system bank 1 rich, and DTC P0174 – Fuel system bank 2 rich. What could cause the condition?

   A. A kinked fuel return hose
   B. A stuck open injector
   C. A cracked MAP sensor vacuum hose
   D. Excessive fuel pressure

Answer A is incorrect. The composite vehicle uses a returnless fuel system.

Answer B is incorrect. A stuck open injector would only affect one bank of the engine and not cause a rich code for both banks.

Answer C is incorrect. The composite vehicle does not have a vacuum hose to the MAP sensor. Furthermore, the MAP sensor is used for diagnostics only, not fuel control.

Answer D is correct. Excessive fuel pressure could cause both banks to run rich.

47. The composite vehicle enters the shop with a DTC P0341 – camshaft position (CMP) sensor performance. Technician A says that this could be caused by an open CMP sensor. Technician B says that this could be caused by excessive camshaft end-play. Who is correct?

TASK C.6

   A. A only

   B. B only

   C. Both A and B

   D. Neither A nor B

   Answer A is incorrect. Typically intermittent problems and out of range readings will set "performance" codes. An open camshaft sensor would set a camshaft sensor circuit open code.

   Answer B is correct. Only Technician B is correct. Excessive camshaft end-play could cause the sensor to reluctor wheel clearance to change, which could alter the output signal intermittently.

   Answer C is incorrect. Only Technician B is correct.

   Answer D is incorrect. Technician B is correct.

48. After sitting for ten minutes, the composite vehicle is hard to start, and fuel pressure measures 32 psi (220 kPa). Once started, the engine runs fine with no DTCs. Which of the following could be the cause?

TASK D.12

   A. Leaking fuel pump check valve

   B. Low fuel pump output

   C. Leaking injector o-ring

   D. A lost cam sensor signal

   Answer A is correct. A leaking fuel pump check valve will allow fuel pressure to bleed off and cause long cranking time when starting after sitting.

   Answer B is incorrect. The minimum allowed fuel pressure for the composite vehicle is 45 psi (319 kPa), and since the vehicle does start after long crank times, the pump is able to make this pressure.

   Answer C is incorrect. A leaking injector o-ring would cause a vacuum leak on the runner, which would not affect cranking times.

   Answer D is incorrect. A lost cam sensor signal may cause extended crank time, but it will also set a DTC, which will illuminate the MIL.

49. The composite vehicle is being diagnosed for a P0302 – cylinder 2 misfire. The #2 ignition coil secondary resistance is 2 KΩ. This indicates:

TASK C.3

   A. A shorted secondary coil

   B. An open secondary coil

   C. A shorted primary coil

   D. A normal coil

   Answer A is correct. The resistance of the secondary coil is lower than specification, which indicates a shorted coil.

   Answer B is incorrect. A resistance of O.L. would indicate an open coil.

   Answer C is incorrect. The resistance measurement indicates that the secondary coil is shorted, not the primary.

   Answer D is incorrect. The specification for secondary coil resistance is 10 KΩ. The 2 KΩ measurement is lower than the specification, which is not normal.

SCAN TOOL DATA

Engine Coolant Temperature (ECT) Sensor 4.3 volts	Intake Air Temperature (IAT) Sensor 2.4 volts	Manifold Absolute Pressure (MAP) Sensor 1.8 volts	Throttle Position Sensor (TPS) 2.4 volts
Engine Speed Sensor (rpm) 1700 rpm	Heated Oxygen Sensor (HO$_2$S) 0.9 volts	Vehicle Speed Sensor (VSS) 55 mph	Battery Voltage (B+) 14.3 volts
	Evaporative Emission Canister Solenoid (EVAP) ON	Torque Converter Clutch (TCC) Solenoid ON	EGR Valve Control Solenoid (EGR) 30 percent
Malfunction Indicator Lamp (MIL) ON	Diagnostic Trouble Codes NONE	Open/Closed Loop CLOSED	Fuel Pump Relay (FP) ON

Measured Ignition Timing °BTDC	Base Timing: 10°	Actual Timing: 11°

TASK D.10

50. The composite vehicle is brought into the shop with the concern of black smoke from the tailpipe. Use the data displayed in the above figure to diagnose the fault.

　A.　A faulty O$_2$ sensor

　B.　A skewed ECT sensor

　C.　A skewed IAT sensor

　D.　A stuck open EVAP purge valve

Answer A is incorrect. Black smoke indicates a rich condition, and the O$_2$ sensor is reading correctly for the rich condition.

Answer B is correct. The ECT sensor is showing an improper reading of 4.3 volts, which equals approximately 0°F (-18°C). It is unlikely that this reading is correct because of the black smoke and the IAT reading approximately 90°F (32°C).

Answer C is incorrect. The IAT voltage is a normal signal that would not cause a rich condition.

Answer D is incorrect. While a stuck open EVAP purge solenoid would cause a rich condition, it is not likely to cause black smoke off idle. The ECT is not reading correctly and is displaying in the data.

PREPARATION EXAM 2—ANSWER KEY

1.	B	21.	D	41.	B
2.	A	22.	C	42.	A
3.	D	23.	D	43.	C
4.	A	24.	B	44.	D
5.	C	25.	A	45.	A
6.	B	26.	C	46.	C
7.	A	27.	B	47.	B
8.	A	28.	C	48.	D
9.	D	29.	D	49.	A
10.	A	30.	A	50.	C
11.	C	31.	D		
12.	A	32.	C		
13.	A	33.	B		
14.	C	34.	A		
15.	D	35.	C		
16.	B	36.	A		
17.	C	37.	C		
18.	D	38.	D		
19.	A	39.	B		
20.	B	40.	A		

PREPARATION EXAM 2—EXPLANATIONS

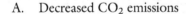

1. Which of the following emissions changes is LEAST LIKELY to be caused by a clogged fuel injector?

    A. Decreased CO_2 emissions

    B. Decreased O_2 emissions

    TASK F.7

    C. Decreased NO_x emissions

    D. Increased CO emissions

    Answer A is incorrect. CO_2 emissions are highest when the engine is operating at a 14.7:1 air-to-fuel ratio. A clogged fuel injector will lower CO_2 emissions.

    Answer B is correct. A clogged fuel injector will cause a lean condition, which will cause high O_2 levels in the exhaust.

    Answer C is incorrect. A clogged fuel injector will not change NO_x emissions.

    Answer D is incorrect. A clogged fuel injector will not increase CO emissions, since it will cause the engine to run lean.

TASK D.16

2. A vehicle was repaired for a DTC P0171 – bank 1 lean. Technician A says that fuel trim percentages can be used to verify the repair. Technician B says that monitoring the O_2 sensor voltages will verify the repair. Who is correct?

   A. A only
   B. B only
   C. Both A and B
   D. Neither A nor B

   Answer A is correct. After a repair of this type, monitoring the fuel trim percentages would be the best way to verify that the vehicle is repaired.

   Answer B is incorrect. While monitoring the O_2 sensor voltage may provide some information on the rich or lean status of the engine, fuel trim percentages are based on many inputs and conditions. The technician cannot accurately determine the rich or lean status by monitoring O_2 sensor voltage alone.

   Answer C is incorrect. Only Technician A is correct.

   Answer D is correct. Only Technician A is correct.

TASK B.2

3. A resistor has been installed in series with the intake air temperature (IAT) sensor in an attempt to increase engine output. This will have what affect on the IAT reading?

   A. Decreased fuel delivery
   B. Decreased signal voltage
   C. Increased temperature reading
   D. Increased signal voltage

   Answer A is incorrect. A higher IAT sensor voltage will cause the ECM to command a richer fuel mixture.

   Answer B is incorrect. A resistor will cause sensor signal voltage to increase.

   Answer C is incorrect. A higher IAT sensor voltage will indicate colder air temperatures.

   Answer D is correct. A resistor in series with the IAT sensor will cause sensor signal voltage to increase.

TASK E.16

4. A technician is testing the integrity of the EVAP system after a major fuel system repair. With the smoke tester connected and turned on, the indicator pellet in the flow gauge remains at the top of the flow gauge. This indicates:

   A. A large leak in the system
   B. A small leak in the system
   C. No leak in the system
   D. Nothing

   Answer A is correct. An indicator bullet that remains at the top of the gauge indicates that there is a large leak present. All of the gas that is being pumped into the system is leaking out. The higher the bullet is in the gauge, the larger the leak.

   Answer B is incorrect. If a small leak were present in the system, the indicator bullet would not be at the top of the flow gauge.

   Answer C is incorrect. The indicator bullet would drop to the bottom of the flow gauge after a few seconds if there were no leaks in the system.

   Answer D is incorrect. The bullet at the top of the gauge indicates a large leak in the system. The technician should look for the smoke to locate the leak.

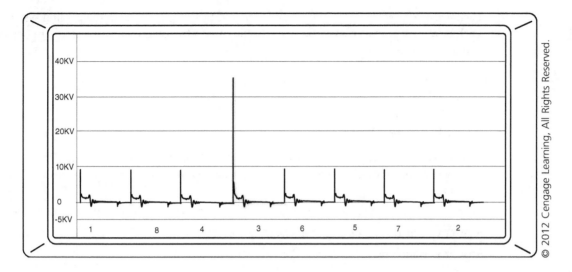

5. The pattern in the above figure is displayed while testing secondary voltage. This condition will cause:

   TASK C.10

   A.   Low O_2 emissions

   B.   Low NO_x emissions

   C.   Excessive HC emissions

   D.   Excessive CO emissions

   Answer A is incorrect. O_2 will not be used in combustion for cylinder #3; therefore, it will cause O_2 levels to rise.

   Answer B is incorrect. The high firing line and short spark line indicate that cylinder #3 is misfiring. An engine misfire will not lower NO_x levels.

   Answer C is correct. A misfiring engine will prevent the HCs and O_2 from burning, which will increase these emissions.

   Answer D is incorrect. Combustion must take place to form CO. No combustion is taking place in cylinder #3.

6. Black smoke is coming from the exhaust under all engine conditions. Technician A says that a throttle position sensor (TPS) that is fixed around 4.5 volts could cause this. Technician B says that an engine coolant temperature (ECT) sensor that is fixed around 4.5 volts could also cause this. Who is correct?

   TASK B.8

   A.   A only

   B.   B only

   C.   Both A and B

   D.   Neither A nor B

   Answer A is incorrect. A TPS that is fixed at 4.5 volts will most likely cause the ECM to enter clear flood mode while cranking.

   Answer B is correct. Only Technician B is correct. An engine coolant temperature sensor that reads 4.5 volts equates to approximately −40° F (−40°C). Based on this temperature, the ECM would over-fuel the engine, which would cause the black smoke.

   Answer C is incorrect. Only Technician B is correct.

   Answer D is incorrect. Technician B is correct.

TASK C.8

7. Technician A says that spark knock can be caused by excessive scale in the cooling system. Technician B says that spark knock can also be caused by using fuel with a too-high octane rating. Who is correct?

   A. A only
   B. B only
   C. Both A and B
   D. Neither A nor B

   Answer A is correct. Only Technician A is correct. Excessive scale buildup in the engine's cooling system will cause combustion chamber temperatures to increase, which can cause spark knock.

   Answer B is incorrect. Using a fuel with a too-low octane rating can lead to spark knock.

   Answer C is incorrect. Only Technician A is correct.

   Answer D is incorrect. Technician A is correct.

TASK A.10

8. A vehicle is brought into the shop with a poor fuel economy concern. Technician A says that installing larger wheels and tires can throw the odometer off and cause this concern. Technician B says that installing a higher-ratio final drive gearset can cause this concern. Who is correct?

   A. A only
   B. B only
   C. Both A and B
   D. Neither A nor B

   Answer A is correct. Only Technician A is correct. Installing larger wheels and tires can cause the odometer to read fewer miles than were actually traveled and can lead to fuel economy complaints.

   Answer B is incorrect. Installing a higher-ratio final drive gearset would allow the vehicle to achieve better fuel economy, since the vehicle would be able to cruise at a lower RPM.

   Answer C is incorrect. Only Technician A is correct.

   Answer D is incorrect. Technician A is correct.

TASK B.19

9. An O_2 sensor is fixed lean. Which of the following can be performed to change O_2 voltage and confirm the operation of the sensor?

   A. Create a vacuum leak
   B. Unplug an injector
   C. Remove spark from a cylinder
   D. Open the EVAP purge valve

   Answer A is incorrect. A vacuum leak would create a lean condition, which would not cause the sensor voltage to change.

   Answer B is incorrect. Unplugging an injector would prevent the oxygen in the cylinder from burning, which would cause the higher O_2 levels in the exhaust and no change in O_2 voltage levels if it were fixed low.

   Answer C is incorrect. Removing spark from the cylinder would cause an excess of unburned oxygen in the cylinder, which would not change the voltage of a lean-reading O_2 sensor.

   Answer D is correct. Opening the EVAP purge valve would create a rich condition, which would cause a functioning O_2 sensor to show the rich condition by creating a higher voltage.

10. The EGR system is being tested after cleaning clogged passages. What should result when the valve is commanded open at idle with the scan tool?

TASK E.16

   A. Rough unstable idle

   B. Increase in idle speed

   C. Black smoke from the exhaust

   D. Low O_2 sensor voltage

   Answer A is correct. A functioning EGR system will cause a rough unstable idle due to the lack of oxygen in the cylinder when commanded open at idle.

   Answer B is incorrect. Opening the EGR valve at idle will not cause an increase in idle speed. The engine should stall if the EGR valve is opened at idle.

   Answer C is incorrect. Opening the EGR valve at idle will not cause black smoke to come out of the exhaust. The exhaust gas should cause an engine stall if allowed to enter the intake at idle.

   Answer D is incorrect. Opening the EGR valve at idle will displace oxygen, and the lack of oxygen in the exhaust would cause the O_2 sensor voltage to increase.

11. A vehicle has come into the shop with an intermittent DTC P0340 – camshaft sensor circuit malfunction. Technician A says that this may be caused by a faulty cam sensor. Technician B says that the cam sensor can be damaged by excessive cam thrust. Who is correct?

TASK A.7

   A. A only

   B. B only

   C. Both A and B

   D. Neither A nor B

   Answer A is incorrect. Technician B is also correct.

   Answer B is incorrect. Technician A is also correct.

   Answer C is correct. Both Technicians are correct. An intermittent P0340 could be caused by a faulty cam sensor or by a sensor that has been damaged by excessive thrust of the camshaft.

   Answer D is incorrect. Both Technicians are correct.

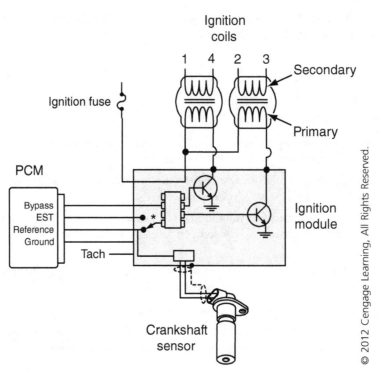

TASK C.4

12. Technician A says that the ignition module in the above image controls the current through the primary ignition coils. Technician B says that the module must have a cam sensor input so that it knows when to turn on each coil. Who is correct?

　　A.　A only
　　B.　B only
　　C.　Both A and B
　　D.　Neither A nor B

Answer A is correct. Only Technician A is correct. The ignition module commands the ignition coils on and controls the current through the primary coil.

Answer B is incorrect. The pictured system is a waste spark system. A waste spark system does not need a cam signal, since the spark plugs of two companion cylinders are fired at one time.

Answer C is incorrect. Only Technician A is correct.

Answer D is incorrect. Technician A is correct.

TASK D.9

13. The throttle of a vehicle with throttle actuator control (TAC) is unresponsive, and idle speed is high. Technician A says that there may not be any power to the throttle motor. Technician B says that one faulty TPS could cause this condition. Who is correct?

　　A.　A only
　　B.　B only
　　C.　Both A and B
　　D.　Neither A nor B

Answer A is correct. Only Technician A is correct. Typically, if the TAC system is inoperative from a fault such as no power, a spring will hold the throttle open around 15 percent and cause the idle speed to be raised. This will allow the engine to run, but the vehicle will not be drivable.

Answer B is incorrect. Only one faulty TPS will cause most TAC systems to go into a limited mode of operation. This will limit throttle operation, but will not shut it totally off.

Answer C is incorrect. Technician A is correct.

Answer D is incorrect. Only Technician A is correct.

14. Technician A says that the fuel level sensor must be functional before the EVAP system can be tested. Technician B says that the O_2 sensor monitor must run and pass before the catalytic converter monitor will run. Who is correct?

    A. A only

    B. B only

    C. Both A and B

    D. Neither A nor B

TASK B.7

    Answer A is incorrect. Technician B is also correct.

    Answer B is incorrect. Technician A is also correct.

    Answer C is correct. Both Technicians are correct. Both technicians are referring to enabling criteria for certain monitors. To run the EVAP system monitor, the ECM needs to know the fuel level. If the fuel level sensor is faulty, the test will be aborted. To run the catalyst monitor, the O_2 sensors must be good. If the O_2 sensors fail their monitors, the catalyst monitor is aborted.

    Answer D is incorrect. Both Technicians are correct.

15. A monitor readiness status of "Complete" indicates:

    A. The vehicle will pass an I/M test

    B. The monitor has passed

    C. The monitor has failed

    D. The monitor has run

TASK F.14

    Answer A is incorrect. There are more tests the vehicle must pass to complete the I/M test.

    Answer B is incorrect. The monitor readiness status cannot be used to determine if the monitor has passed or failed. A technician must look at hard DTCs or pending DTCs to determine if the monitor passed or failed.

    Answer C is incorrect. The monitor readiness status cannot be used to determine if the monitor has passed or failed. DTCs and test results are the only way to determine if the monitor passed or failed.

    Answer D is correct. A monitor readiness status of "complete" only indicates that the monitor has run.

16. Exhaust emissions are being tested on a vehicle. O_2 levels are higher than normal, and CO_2 levels are low. Which of the following would LEAST LIKELY cause this condition?

    A. A restricted injector

    B. A stuck closed PCV valve

    C. A clogged fuel filter

    D. A vacuum leak

TASK A.9

    Answer A is incorrect. A restricted injector would cause a lean condition, which would not burn oxygen in the cylinder.

    Answer B is correct. A stuck closed PCV valve could restrict the amount of airflow into the cylinder, which would create a lack of oxygen.

    Answer C is incorrect. A clogged fuel filter would cause a lean condition, which would increase O_2 levels and lower CO_2 levels in the exhaust.

    Answer D is incorrect. A vacuum leak would cause the engine to run lean, which would show up as more oxygen content in the exhaust. CO_2 levels would also decrease.

TASK F.5

17. Technician A says that it is not possible to have both high O_2 and CO emissions. Technician B says that O_2 is a lean indicator and CO is a rich indicator. Who is correct?

    A. A only

    B. B only

    C. Both A and B

    D. Neither A nor B

Answer A is incorrect. Technician B is also correct.

Answer B is incorrect. Technician A is also correct.

Answer C is correct. Both Technicians are correct. Since high O_2 levels indicate a lean condition and high CO levels indicate a rich condition, the two gases can never be high at the same time during an exhaust emissions test.

Answer D is incorrect. Both Technicians are correct.

TASK B.8

18. What effect would a stuck open EGR valve have on O_2 sensor readings?

    A. Low steady voltage readings

    B. Slowly switching voltage readings

    C. Rapidly switching voltage readings

    D. High steady voltage readings

Answer A is incorrect. Low O_2 voltages are associated with high oxygen content in the exhaust. The exhaust gas would displace the oxygen and cause the O_2 voltages to be high.

Answer B is incorrect. A slowly switching O_2 sensor is typically an indicator of a lazy or failing O_2 sensor

Answer C is incorrect. A rapidly switching O_2 sensor is an indicator of a normally functioning O_2 sensor.

Answer D is correct. An EGR valve that is stuck open will displace oxygen in the combustion chamber. This will cause the oxygen that does get into the cylinder to burn, indicated by a steady high voltage reading.

TASK D.10

19. A vehicle with a return type fuel system is hard to start after sitting and has black smoke from the exhaust after finally starting. This could be caused by:

    A. A leaking fuel pressure regulator

    B. Low fuel pump output

    C. A leaking fuel pump check valve

    D. A plugged fuel filter

Answer A is correct. A leaking fuel pressure regulator can allow fuel pressure to bleed off into the intake, which will allow too much fuel into the intake upon starting. The rich mixture and lack of a primed fuel system will cause the vehicle to be hard to start.

Answer B is incorrect. While low fuel pump output will cause the vehicle to be hard to start, it will not cause black smoke on start up.

Answer C is incorrect. A leaking fuel pump check valve will cause hard starting; it will not cause black smoke upon start-up.

Answer D is incorrect. A plugged fuel filter will not cause black smoke on start up, as black smoke is caused by a rich condition, and a plugged fuel filter would cause a lean condition.

20. The ignition module on a distributor ignition (DI) vehicle has been replaced several times after it has failed. Technician A says that this can be caused by a shorted magnetic pick-up. Technician B says that this can also be caused by a shorted ignition coil. Who is correct?

TASK C.13

   A. A only
   B. B only
   C. Both A and B
   D. Neither A nor B

   Answer A is incorrect. A shorted magnetic pick-up may cause the vehicle not to start, but it would not cause damage to the ignition module.

   Answer B is correct. Only Technician B is correct. A shorted ignition coil would cause excessive current flow through the ignition module, which could cause it to fail. If the coil is not replaced, the same failure will reoccur.

   Answer C is incorrect. Only Technician B is correct.

   Answer D is incorrect. Technician B is correct.

Ohmmeter
(no continuity)

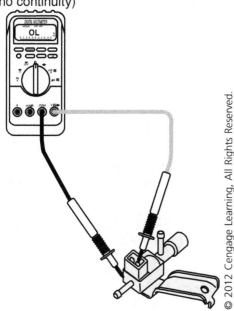

© 2012 Cengage Learning, All Rights Reserved.

21. The EVAP purge solenoid in the above figure is being tested. Technician A says that the solenoid is open. Technician B says that the solenoid must be replaced. Who is correct?

TASK E.9

   A. A only
   B. B only
   C. Both A and B
   D. Neither A nor B

   Answer A is incorrect. The solenoid should not have continuity to the outer housing, so this reading is normal. The actual resistance of the solenoid itself is not being measured.

   Answer B is incorrect. The meter should not read continuity during this test; therefore, the solenoid should not be considered faulty based on the results of this test.

   Answer C is incorrect. Neither Technician is correct.

   Answer D is correct. Neither Technician is correct. A reading of O.L., or no continuity from the housing of the solenoid to the terminal, is a good reading. If a solenoid has any reading other than O.L., the solenoid is bad.

TASK B.5

22. A vehicle has an illuminated MIL and a DTC P0123 – TPS high voltage. Technician A says that an open signal return wire could cause this code. Technician B says that a shorted TPS could cause this DTC. Who is correct?

    A. A only

    B. B only

    C. Both A and B

    D. Neither A nor B

    Answer A is incorrect. Technician B is also correct.

    Answer B is incorrect. Technician A is also correct.

    Answer C is correct. Both Technicians are correct. An open in the signal return circuit could cause the signal circuit to read abnormally high. A short inside the TPS can also cause a high signal voltage.

    Answer D is incorrect. Both Technicians are correct.

TASK F.8

23. A vehicle has failed a no-load I/M test for high HC emissions, but HC emissions are acceptable at higher RPM. Technician A says that this can be caused by a fouled spark plug. Technician B says that this could be caused by low fuel pressure. Who is correct?

    A. A only

    B. B only

    C. Both A and B

    D. Neither A nor B

    Answer A is incorrect. A fouled spark plug would cause HC emissions to be high at all times, including higher RPM.

    Answer B is incorrect. Low fuel pressure may allow the engine to run normally at idle but would cause a lean misfire at higher RPM. The misfire would cause high HC emissions.

    Answer C is incorrect. Neither Technician is correct.

    Answer D is correct. Neither Technician is correct. This condition could likely be caused when a lean misfire is present, such as with a large vacuum leak.

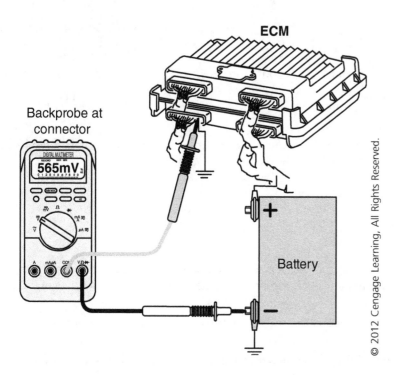

ECM

Backprobe at connector

565mV

Battery

24. A vehicle is brought into the shop with multiple DTCs and numerous drivability concerns. Testing of one of the ECM grounds is displayed above. This test indicates:

    A.   High resistance in the ECM power circuit

    B.   High resistance in the ECM ground circuit

    C.   Continuity of the ground circuit

    D.   The ground circuit is shorted to power

TASK B.14

Answer A is incorrect. The voltage drop of the power circuit is not being measured.

Answer B is correct. The reading on the meter shows that the ground circuit has a high voltage drop, indicating high resistance, and is using 565 mV, which results in the given symptoms.

Answer C is incorrect. Continuity of the ground circuit is not being measured.

Answer D is incorrect. The voltage drop of the ground circuit is being measured. A power wire shorted to a ground circuit would likely cause a fuse to blow.

25. A cracked exhaust manifold will cause O_2 sensor readings to:

    A.   Read low voltage

    B.   Read high voltage

    C.   Switch rapidly

    D.   Be fixed at 450 mV

TASK A.11

Answer A is correct. A cracked exhaust manifold will allow excess oxygen to enter the exhaust stream, which the O_2 sensor would indicate through a low voltage signal.

Answer B is incorrect. A cracked exhaust manifold would cause the O_2 sensor to read lean, which is associated with low voltage. High O_2 voltage is associated with a rich condition.

Answer C is incorrect. An exhaust manifold leak may cause the sensor to be fixed at a low voltage. This would slow the switch rate.

Answer D is incorrect. The excess oxygen would cause the sensor to generate a low voltage signal.

Cylinder	1	2	3	4	5	6
High reading	225	225	225	225	225	225
Low reading	100	100	100	100	115	115
Amount of drop	125	125	125	125	110	110

TASK D.12

26. The results from an injector balance test are displayed in the table above. These results can cause:

A. A rich condition

B. An engine miss

C. A lean condition

D. Low engine power

Answer A is incorrect. Injectors five and six are not flowing as much as the rest, which would cause a lean condition. Rich conditions are caused by excessive amounts of fuel in the engine.

Answer B is incorrect. Injectors five and six are still flowing and would not create a complete miss.

Answer C is correct. Injectors five and six are not flowing as much as the rest of the injectors. This would cause an overall lean condition.

Answer D is incorrect. The ECM could increase injector pulse width (PW) to make up for the lack of fuel. The driver would probably not notice a difference in power output.

TASK E.6

27. An air injection system that constantly pumps air upstream will cause the O_2 sensors to:

A. Read rich

B. Create low voltage

C. Create high voltage

D. Fluctuate rapidly

Answer A is incorrect. Since the air injection system pumps oxygen into the exhaust, the O_2 sensor will not read rich.

Answer B is correct. When air is injected into the exhaust, the oxygen level of the exhaust is raised. This will cause the O_2 sensors to produce a low voltage signal.

Answer C is incorrect. The O_2 sensor will display a lean reading, which will be a low voltage.

Answer D is incorrect. With oxygen being pumped into the exhaust, the O_2 sensor voltage will not fluctuate much at all.

28. An ECT sensor is being checked for proper operation. While the voltage is being monitored on the scan tool after a cold start, what should the voltage do as the sensor heats up?

TASK B.9

   A. Switch rapidly

   B. Steadily increase

   C. Steadily decrease

   D. Slowly fluctuate

   Answer A is incorrect. The ECT sensor voltage should not switch rapidly; it should change slowly along with slight fluctuations in engine temperature.

   Answer B is incorrect. ECT sensor voltage will slowly decrease as the temperature rises.

   Answer C is correct. High voltage readings from the ECT sensor indicate cold operating conditions. As the engine heats up, the voltage will steadily drop.

   Answer D is incorrect. The ECT sensor voltage should not fluctuate; it should change slowly based on the temperature changes of the engine.

29. The scan tool will not communicate with the vehicle's ECM, and the MIL will not illuminate during the bulb test. The most likely cause of this is:

TASK F.15

   A. A bad diagnostic link connection (DLC)

   B. No power to the DLC

   C. A malfunctioning scan tool

   D. An open ECM ground

   Answer A is incorrect. If there were a bad connection at the DLC, the MIL would still illuminate during the bulb check.

   Answer B is incorrect. If there were no power at the DLC, the MIL would still illuminate during the bulb check, and the scan tool would not power up.

   Answer C is incorrect. If the scan tool were malfunctioning, the MIL would still illuminate during the bulb check; however, this can easily be tested by connecting the scan tool to another vehicle

   Answer D is correct. If the ECM ground circuit were open, the ECM would not power up and would have no communication with the scan tool.

30. A vehicle has a high idle and no A/C. Which of the following could cause both of these symptoms?

TASK B.11

   A. A shorted power steering pressure switch

   B. An intake manifold vacuum leak

   C. A hanging throttle plate

   D. Higher-than-normal TPS voltage

   Answer A is correct. A shorted power steering pressure switch would cause the ECM to believe that power steering pressure is high. The ECM would then increase idle speed to handle the extra engine load and disable the A/C.

   Answer B is incorrect. While an intake manifold vacuum leak may cause a higher-than-normal idle speed, it will not cause the A/C to be inoperative.

   Answer C is incorrect. While a hanging throttle plate may cause higher-than-normal idle speed, it would not cause the A/C to be inoperative.

   Answer D is incorrect. Higher-than-normal TPS voltage may cause the A/C not to function under certain conditions, but it would not cause a higher-than-normal idle speed.

TASK E.9

31. The outlet temperature of the catalytic converter is 910°F (488°C) and the inlet is 1005°F (541°C). This indicates:

    A. The converter is overheating

    B. There is a misfire

    C. A degraded converter

    D. Normal operation

    Answer A is incorrect. This increase in temperature does not indicate an overheating converter. The converter will typically be able to stand temperatures up to 1600°F (871°C).

    Answer B is incorrect. A misfire will cause the temperature rise to be more than 10 percent.

    Answer C is correct. If the catalytic converter were degraded, there would be no temperature rise between the inlet and outlet of the catalytic converter.

    Answer D is incorrect. A properly functioning catalytic converter will create heat as it converts HC into H_2O and CO_2. The outlet should be approximately 10 percent hotter than the inlet.

TASK F.9

32. Which of the following valves, if stuck open, would cause highest increase in CO emissions?

    A. EGR

    B. EVAP vent

    C. EVAP purge

    D. Diverter

    Answer A is incorrect. A stuck open EGR valve would cause an increase in HC emissions due to the lack of oxygen. CO may only slightly increase.

    Answer B is incorrect. The EVAP vent is normally open and will not affect CO emissions.

    Answer C is correct. A stuck open EVAP purge valve would cause a rich condition, which would increase CO emissions.

    Answer D is incorrect. A diverter valve has no real open or closed position. It directs air either upstream or downstream into the exhaust system.

TASK A.11

33. A vehicle arrives in the shop with a DTC P0401 – EGR insufficient flow. All of the following could cause this EXCEPT:

    A. Aftermarket exhaust system

    B. Restricted catalytic converter

    C. Broken EGR valve vacuum hose

    D. Plugged EGR passages

    Answer A is incorrect. A low backpressure aftermarket exhaust system could cause the EGR system to function improperly.

    Answer B is correct. A restricted catalytic converter will not cause a lack of EGR flow. It could actually cause excessive EGR flow.

    Answer C is incorrect. A broken vacuum hose to the EGR valve would cause the system to be inoperative and set a DTC P0401.

    Answer D is incorrect. Plugged EGR passages would cause insufficient EGR flow.

34. A stuck open EVAP purge solenoid would cause:

    A. Increased CO emissions

    B. Increased CO_2 emissions

    C. Decreased NO_x emissions

    D. Decreased HC emissions

    TASK E.10

    Answer A is correct. An EVAP purge solenoid would cause a rich condition, which would raise CO emissions.

    Answer B is incorrect. CO_2 emissions are highest when the air/fuel ratio is at 14.7:1. A leaking purge solenoid would richen that ratio.

    Answer C is incorrect. NO_x emissions are lowered by the EGR valve.

    Answer D is incorrect. With a rich condition, HC emissions are likely to increase instead of decrease.

35. A vehicle with a return type fuel system has failed an emissions test with the following readings at idle:

    TASK F.7

    HC – 20 ppm

    CO – 3 percent

    CO_2 – 10 percent

    O_2 – 0 percent

    Which of the following could cause these readings?

    A. An intake manifold vacuum leak

    B. A plugged fuel filter

    C. A leaking fuel pressure regulator

    D. A plugged EGR passage

    Answer A is incorrect. An intake manifold vacuum leak would cause O_2 emissions to increase.

    Answer B is incorrect. A plugged fuel filter would cause a lean condition, and O_2 levels would increase.

    Answer C is correct. A leaking fuel pressure regulator will cause a rich condition, which will increase CO emissions.

    Answer D is incorrect. The EGR system controls NO_x emissions and is not operational at idle.

TASK B.13

36. Technician A says that most immobilizer systems will turn off the fuel injectors if there is a fault in the anti-theft system. Technician B says that the ignition system is also disabled. Who is correct?

  A. A only
  B. B only
  C. Both A and B
  D. Neither A nor B

Answer A is correct. Only Technician A is correct. When there is a fault in the vehicle immobilizer system, the ECM will turn off the fuel injectors immediately after start up.

Answer B is incorrect. If the ignition system were disabled when there was an immobilizer fault, it may be possible for unburned HCs to enter the catalytic converter. For this reason, the ignition system will stay operational if there is an immobilizer system fault.

Answer C is incorrect. Only Technician A is correct.

Answer D is incorrect. Technician A is correct.

TASK D.12

37. A fuel injector can be tested for all of the following EXCEPT:

  A. Flow
  B. Resistance
  C. Pressure
  D. Current draw

Answer A is incorrect. The fuel injector can be tested for flow by performing an injector balance test.

Answer B is incorrect. The resistance of a fuel injector can be tested and compared to spec.

Answer C is correct. A fuel injector does not make pressure, and it cannot be tested.

Answer D is incorrect. The current draw of a fuel injector can be checked and used to determine if the injector is good.

TASK E.10

38. Engine detonation can be caused by:

  A. An inoperative EVAP purge valve
  B. A clogged catalytic converter
  C. A clogged PCV valve
  D. An inoperative EGR valve

Answer A is incorrect. An inoperative EVAP purge valve is likely to cause high evaporative emissions but will not cause detonation.

Answer B is incorrect. A clogged catalytic converter is likely to cause poor engine power but will not cause detonation.

Answer C is incorrect. A clogged PCV valve is likely to cause high emissions or oil leakage but will not cause detonation.

Answer D is correct. A non-functioning EGR valve will allow cylinder temperatures to increase. High temperature conditions can cause detonation.

39. The composite vehicle is brought into the shop with a DTC P0302 – cylinder #2 misfire. The cause of this misfire could be:

TASK C.3

    A. A blown fuse #4

    B. An open wire from the coil to terminal 17 of the ECM

    C. An open wire from the coil to terminal 18 of the ECM

    D. High secondary plug wire resistance

    Answer A is incorrect. If fuse #4 were blown, the vehicle would not start.

    Answer B is correct. Terminal 17 of the ECM provides ground for the #2 ignition coil. If the wire from the coil to this terminal were open, the coil would not fire.

    Answer C is incorrect. Terminal 18 of the ECM provides ground for cylinder #4's ignition coil.

    Answer D is incorrect. The composite vehicle does not use spark plug wires.

40. The ECM has been replaced several times because the A/C clutch relay driver has failed. This could be caused by:

TASK B.23

    A. A shorted relay coil

    B. High relay coil resistance

    C. High A/C clutch coil resistance

    D. A shorted A/C clutch coil

    Answer A is correct. A shorted relay coil would cause high current to flow through the ECM, which would cause damage to the relay driver. If the relay is not replaced, the driver will fail again.

    Answer B is incorrect. High resistance in the relay coil circuit will not damage the ECM, since current flow will be lower than normal.

    Answer C is incorrect. High resistance in the clutch coil circuit will not damage the ECM, since it is not directly connected to the ECM.

    Answer D is incorrect. A shorted A/C clutch coil may cause a fuse to blow if current flow were high enough. The current would not damage the ECM, since it flows through the relay and not the ECM.

41. The composite vehicle is slow to fill when refueling. This could be due to all of the following EXCEPT:

TASK E.12

    A. A pinched hose from the vent solenoid to the intake

    B. A stuck closed purge solenoid

    C. A pinched hose from the tank to the EVAP canister

    D. A stuck closed vent solenoid

    Answer A is incorrect. A pinched hose from the vent solenoid to the intake would not allow air to leave the tank while refueling and could cause the tank to be slow to fill.

    Answer B is correct. The purge solenoid is normally closed and would not affect the filling of the fuel tank if it were stuck closed.

    Answer C is incorrect. A pinched hose from the tank to the EVAP canister would not allow air to leave the tank while refueling and would cause the tank to be slow to fill.

    Answer D is incorrect. A stuck closed vent solenoid would not allow air to leave the tank while refueling and would cause the tank to be slow to fill.

TASK A.1

42. The composite vehicle engine speed will not rise above 6000 rpm. This is most likely due to:

A.   ECM programming

B.   A restricted catalytic converter

C.   Low fuel pump output pressure

D.   A slipped timing chain

Answer A is correct. The ECM will not allow the engine RPM to rise above 6000 rpm for safety reasons.

Answer B is incorrect. A restricted catalytic converter would limit engine RPM well below 6000 rpm.

Answer C is incorrect. It is not likely that the engine would reach 6000 rpm if the fuel pump had low output.

Answer D is incorrect. It is not likely for a timing chain to slip.

TASK D.4

43. The composite vehicle has been brought into the shop with the MIL illuminated, a DTC P0172 – bank 1 rich, and P0174 – bank 2 rich. The cause for this condition could be:

A.   A leaking fuel injector

B.   A leaking fuel pressure regulator

C.   A short to ground at ECM terminal 110

D.   A short to ground at ECM terminal 120

Answer A is incorrect. A leaking fuel injector would cause a rich condition for only one bank.

Answer B is incorrect. The composite vehicle's fuel pressure regulator is in the fuel tank.

Answer C is correct. Terminal 110 of the ECM grounds the EVAP purge solenoid. If this terminal were grounded, the purge solenoid would open and allow HC vapors to get into the intake. These HC vapors would cause a rich condition.

Answer D is incorrect. A short to ground at ECM terminal 120 would cause cylinder #1's injector to stay on. This condition would only affect one bank.

SCAN TOOL DATA

Engine Coolant Temperature Sensor (ECT) 0.35 Volts	Intake Air Temperature Sensor (IAT) 4.50 Volts	MAP Sensor 4.5 V MAF Sensor 0.2 V	Throttle Position Sensor (TPS) 0.6 Volts
Engine Speed Sensor (RPM) 0 rpm	Heated O$_2$ HO$_2$S Upstream 0 V Downstream 0 V	Vehicle Speed Sensor (VSS) 0 mph	Battery Voltage (B+) 12.6 Volts
	Evaporative Emission Canister Solenoid (EVAP) OFF	Torque Converter Clutch Solenoid (TCC) OFF	EGR Valve Control Solenoid (EGR) 0 percent
Malfunction Indicator Lamp (MIL) ON	Diagnostic Trouble Codes NONE	Open/Closed Loop OPEN	Fuel Pump Relay (FP) OFF
Fuel Level Sensor 3.5 V	Fuel Tank Pressure Sensor 2.5 V	Transmission Fluid Temperature Sensor 3.5 V	Transmission Turbine Shaft Speed Sensor 0 mph
Transmission Range Switch PARK	Trans Pressure Control Solenoid 0	Transmission Shift Solenoid 1 ON	Transmission Shift Solenoid 2 OFF

Measured Ignition Timing °BTDC   Base Timing: 10°   Actual Timing: 20°

44. The composite vehicle has been brought into the shop because of poor fuel economy. Based on the key on and the engine off (KOEO) scan tool readings shown in the figure above, what could be the cause?

TASK B.8

A. A faulty ECT sensor

B. A cracked MAP sensor vacuum line

C. A stuck closed EGR valve

D. A faulty IAT sensor

Answer A is incorrect. The ECT sensor is reading around 212°F (100°C), which would not cause poor fuel economy.

Answer B is incorrect. Since the data is being read KOEO, it cannot be determined if the vacuum hose is cracked.

Answer C is incorrect. The EGR valve should be closed under these conditions.

Answer D is correct. The IAT sensor is reading approximately –10°F (–23°C). This could cause the ECM to richen the air/fuel mixture, which would decrease fuel economy.

45. The composite vehicle will not start. What is the minimum voltage required while measuring the output of the crankshaft position sensor while cranking?

TASK C.4

A. 5.0 VAC

B. 2.0 VDC

C. 0.5 VDC

D. 0.2 VAC

Answer A is correct. The ECM must see a minimum of 5.0 VAC from the crank sensor or else the engine will not start.

Answer B is incorrect. The crank sensor produces AC voltage.

Answer C is incorrect. The crank sensor produces AC voltage.

Answer D is incorrect. 0.2 VAC is not enough voltage for the engine to start.

TASK B.10

46. A DTC P0102 – MAF sensor low voltage could be caused by all of the following EXCEPT:

A. A clogged exhaust

B. Air duct leaks

C. Water intrusion

D. A vacuum leak

Answer A is incorrect. A clogged exhaust will restrict airflow, which would cause the low voltage signal.

Answer B is incorrect. Air duct leaks will allow unmetered air to enter the engine, which would cause the low MAF sensor voltage.

Answer C is correct. Water entering the air stream will cool the heated element of the MAF sensor. This will cause the sensor to read a higher voltage, indicating more airflow.

Answer D is incorrect. A vacuum leak will allow unmetered air to enter the engine, which would cause the low MAF sensor voltage.

TASK D.10

47. The composite vehicle is brought into the shop with a spark knock concern. What should the technician do?

A. Adjust ignition timing

B. Fill the tank with quality fuel

C. Check for excessive fuel pressure

D. Adjust the TPS

Answer A is incorrect. The ignition timing of the composite vehicle cannot be adjusted.

Answer B is correct. Low octane fuel can cause the engine to spark knock. To test for this, the tank can be filled with known good fuel.

Answer C is incorrect. Excessive fuel pressure would cause a rich condition, which would not be associated with a spark knock condition.

Answer D is incorrect. The TPS of the composite vehicle is not adjustable.

TASK B.3

48. The composite vehicle is brought into the shop with a DTC P0031 - AFR Sensor heater circuit. Technician A says that this could be caused by a blown fuse #20. Technician B says that this could be caused by an open in the wire between terminal B of the AFR sensor and terminal 151 of the ECM. Who is correct?

A. A only

B. B only

C. Both A and B

D. Neither A nor B

Answer A is incorrect. Fuse #20 powers many different components, not just the HO_2S 1/1 heater. If this fuse were blown, there would be many more codes set.

Answer B is incorrect. Terminal 151 of the ECM is part of the AFR signal circuit. An open in this wire would not set a heater circuit DTC.

Answer C is incorrect. Neither Technician is correct.

Answer D is correct. Neither Technician is correct. This code is typically caused by an open AFR sensor heater element.

49. The composite vehicle has a rough idle but drives normally at cruise. Technician A says that terminal 108 at the ECM could be grounded. Technician B says that a grounded ECM terminal 104 can also cause this symptom. Who is correct?

TASK E.3

   A. A only

   B. B only

   C. Both A and B

   D. Neither A nor B

   Answer A is correct. Only Technician A is correct. Terminal 108 at the ECM controls the EGR valve. If this terminal were shorted to ground, the EGR valve would be open at all times.

   Answer B is incorrect. ECM terminal 104 controls the EVAP vent solenoid. If the vent solenoid were on at all times, the vehicle's idle would not be affected.

   Answer C is incorrect. Only Technician A is correct.

   Answer D is incorrect. Technician A is correct.

50. The composite vehicle has failed a plug in emissions test because the catalytic converter monitor displayed "Not Complete." Technician A says that this means the monitor has not run. Technician B says that the monitor may not have run because one of the AFR sensor monitors may have failed. Who is correct?

TASK F.14

   A. A only

   B. B only

   C. Both A and B

   D. Neither A nor B

   Answer A is incorrect. Technician B is also correct.

   Answer B is incorrect. Technician A is also correct.

   Answer C is correct. Both Technicians are correct. The catalyst monitor for the composite vehicle will only run if both the AFR sensor and the AFR sensor heater monitors have run and passed.

   Answer D is incorrect. Both Technicians are correct.

PREPARATION EXAM 3—ANSWER KEY

1.	A	21.	C	41.	A
2.	D	22.	D	42.	C
3.	A	23.	B	43.	C
4.	C	24.	A	44.	A
5.	B	25.	D	45.	B
6.	A	26.	B	46.	D
7.	D	27.	A	47.	B
8.	C	28.	C	48.	C
9.	B	29.	B	49.	D
10.	A	30.	D	50.	B
11.	D	31.	C		
12.	A	32.	D		
13.	C	33.	D		
14.	C	34.	A		
15.	A	35.	A		
16.	B	36.	D		
17.	D	37.	A		
18.	B	38.	B		
19.	D	39.	B		
20.	C	40.	D		

PREPARATION EXAM 3—EXPLANATIONS

TASK A.9

1. Technician A says that excessive combustion chamber deposits could cause high NO_x emissions. Technician B says that excessive combustion chamber deposits could also cause excessive CO_2 emissions. Who is correct?

    A. A only
    B. B only
    C. Both A and B
    D. Neither A nor B

    Answer A is correct. Only Technician A is correct. Excessive combustion chamber deposits can increase combustion chamber pressure and temperature. It is under these conditions that NO_x is formed.

    Answer B is incorrect. Excessive combustion chamber deposits will not cause excessive CO_2 emissions.

    Answer C is incorrect. Only Technician A is correct.

    Answer D is incorrect. Technician A is correct.

2. While testing the function of the exhaust gas recirculation (EGR) system, what should be noticed when the EGR valve is opened with the key on and the engine running (KOER)?

TASK E.9

A. Increased engine speed

B. Black exhaust smoke

C. Lowered O_2 sensor readings

D. Rough or unstable idle

Answer A is incorrect. Engine speed will decrease if the EGR valve is opened with the engine running.

Answer B is incorrect. Black smoke is caused by a rich condition and excessive fuel.

Answer C is incorrect. O_2 sensors will read high voltage when the EGR valve is open due to the lack of oxygen.

Answer D is correct. With the engine at idle and the EGR valve open, the idle should become rough and unstable.

3. A throttle position sensor (TPS) is suspected of causing a hesitation. Which of the following tools would LEAST LIKELY be used to test the TPS?

TASK B.17

A. Test light

B. Voltmeter

C. Oscilloscope

D. Ohmmeter

Answer A is correct. A test light cannot and should never be used to test a TPS.

Answer B is incorrect. A voltmeter can be used to test the signal voltage of the TPS while opening the throttle and monitoring its voltage.

Answer C is incorrect. The oscilloscope would best test the TPS for a glitch. If the TPS has an open spot, it would be clearly shown on the scope.

Answer D is incorrect. An ohmmeter can be used to test a TPS for resistance while opening the throttle and watching for an erratic reading.

4. Technician A says that a misfiring cylinder due to a fouled spark plug will increase HC emissions. Technician B says that the fouled plug will increase O_2 levels. Who is correct?

TASK F.4

A. A only

B. B only

C. Both A and B

D. Neither A nor B

Answer A is incorrect. Technician B is also correct.

Answer B is incorrect. Technician A is also correct.

Answer C is correct. Both Technicians are correct. A fouled spark plug will cause HCs and O_2 to leave the cylinder unburned, increasing those levels.

Answer D is incorrect. Both Technicians are correct.

5. The main sensor used for fuel control when the vehicle is in closed loop is:

    A. TPS

    B. O_2

    C. ECT

    D. MAP

    Answer A is incorrect. The TPS is used primarily to determine engine load when the vehicle is in closed loop fuel control.

    Answer B is correct. The O_2 sensor is the main sensor used to determine fuel control when the vehicle is in closed loop.

    Answer C is incorrect. While the ECT sensor is important to fuel delivery, it is considered a low authority sensor while the engine is operating in closed loop fuel control.

    Answer D is incorrect. The MAP sensor is used for engine load and rationality testing when in closed loop.

6. Technician A says that electronic control module (ECM) reprogramming can be done to correct a problem when instructed by a TSB. Technician B says ECM reprogramming can be done during routine maintenance to prevent problems. Who is correct?

    A. A only

    B. B only

    C. Both A and B

    D. Neither A nor B

    Answer A is correct. Only Technician A is correct. An ECM can be reprogrammed to correct a potential software problem when instructed by a TSB.

    Answer B is incorrect. ECM reprogramming should only be performed when required by a recall, TSB, or other diagnostic troubleshooting materials.

    Answer C is incorrect. Only Technician A is correct.

    Answer D is incorrect. Technician A is correct.

7. Technician A says a cracked spark plug will cause firing voltage to increase. Technician B says that a fouled spark plug will also cause firing voltage to increase. Who is correct?

    A. A only

    B. B only

    C. Both A and B

    D. Neither A nor B

    Answer A is incorrect. A cracked spark plug will decrease the total amount of secondary resistance. Decreased resistance will cause the circuit to have a lower firing voltage.

    Answer B is incorrect. A fouled spark plug will typically decrease the resistance of the secondary circuit, which will cause the required voltage to decrease as deposits on the plug make a conductive path to ground, which the spark travels along.

    Answer C is incorrect. Neither Technician is correct.

    Answer D is correct. Neither Technician is correct. Typically, high resistance in secondary circuits will cause the firing voltage to increase.

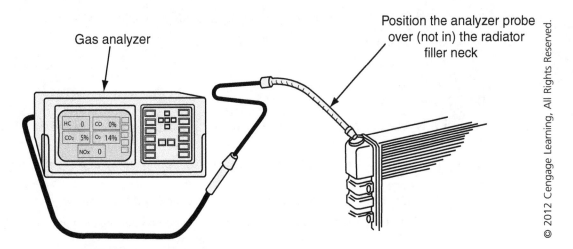

Gas analyzer

Position the analyzer probe over (not in) the radiator filler neck

8. What procedure is being done in the above picture?

    A. Testing engine coolant level
    B. Measuring coolant freeze protection
    C. Checking for a blown head gasket
    D. Calibrating the exhaust gas analyzer

TASK A.8

Answer A is incorrect. Coolant level cannot be tested using the exhaust gas analyzer.

Answer B is incorrect. The exhaust gas analyzer cannot measure coolant freeze protection.

Answer C is correct. The exhaust gas analyzer is being used to check for a blown head gasket. If the analyzer shows any signs of HC emissions from the coolant, a blown head gasket should be suspected.

Answer D is incorrect. The exhaust gas analyzer is not calibrated in the pictured method.

9. A vehicle with a return-type fuel system idles roughly and blows black smoke after a hot soak. Technician A says that this could be caused by low fuel pressure. Technician B says that a leaking fuel pressure regulator can cause this condition. Who is correct?

    A. A only
    B. B only
    C. Both A and B
    D. Neither A nor B

TASK D.10

Answer A is incorrect. Low fuel pressure will not cause black smoke, since it will not cause a rich condition.

Answer B is correct. Technician B is correct. A leaking fuel pressure regulator will allow fuel to leak into the intake manifold. This will cause a rich condition on start up after sitting for a few minutes.

Answer C is incorrect. Only Technician B is correct.

Answer D is incorrect. Technician B is correct.

TASK F.7

10. What condition should be tested for when O_2 levels are high and CO_2 levels are low?

    A.   A lean condition

    B.   A rich condition

    C.   Low idle speed

    D.   High fuel pressure

Answer A is correct. When the engine operates lean, O_2 levels will increase and CO_2 levels will decrease.

Answer B is incorrect. A rich condition will cause O_2 levels to be low and CO levels to be high.

Answer C is incorrect. Low idle speed should not cause increased O_2 levels.

Answer D is incorrect. High fuel pressure would cause a rich condition and increased CO levels.

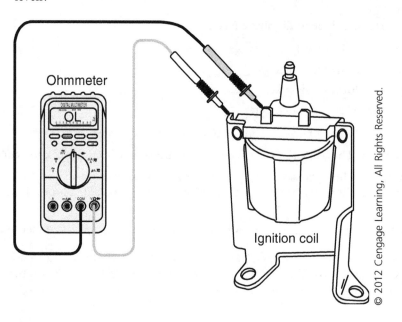

Ohmmeter

Ignition coil

TASK C.5

11. What is being tested in the above picture?

    A.   Secondary coil resistance

    B.   Primary coil resistance

    C.   Secondary coil voltage

    D.   Primary coil short

Answer A is incorrect. The meter is not on the secondary coil terminal, so secondary resistance cannot be tested in this way.

Answer B is incorrect. While one meter lead is on the primary side, the other is on the coil case and testing for a short to ground.

Answer C is incorrect. The meter is in the Ohms position and will not measure voltage in this position.

Answer D is correct. The illustration shows the meter being used to test the resistance from the coil primary winding to the coil case. This action tests for a primary coil short to ground.

12. The ECM has been replaced on a vehicle. Technician A says that the anti-theft system may have to be reprogrammed to recognize the coded ignition key. Technician B says that it is only necessary to program one key. Who is correct?

TASK B.13

A. A only
B. B only
C. Both A and B
D. Neither A nor B

Answer A is correct. Technician A is correct. On many vehicles, when replacing the ECM it will be necessary to program the anti-theft system to recognize the coded ignition keys.

Answer B is incorrect. When programming ignition keys after ECM replacement, it will be necessary to program all keys that will be used for that particular vehicle.

Answer C is incorrect. Only Technician A is correct.

Answer D is incorrect. Technician A is correct.

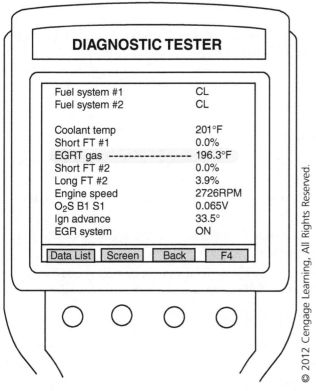

DIAGNOSTIC TESTER

Fuel system #1	CL
Fuel system #2	CL
Coolant temp	201°F
Short FT #1	0.0%
EGRT gas -----------------	196.3°F
Short FT #2	0.0%
Long FT #2	3.9%
Engine speed	2726RPM
O₂S B1 S1	0.065V
Ign advance	33.5°
EGR system	ON

Data List Screen Back F4

13. The EGR system is being tested on a vehicle. The above data is displayed when the EGR valve is actuated with the scan tool. This is an indication of:

TASK F.6

A. EGR flow to one bank only
B. Too much EGR flow
C. Insufficient EGR flow
D. An overheated EGR valve

Answer A is incorrect. There is no evidence of EGR flow to one bank.

Answer B is incorrect. The data shows no EGR flow.

Answer C is correct. The EGR temperature sensor is reading close to the same as the operating temperature of the engine. This indicates that no EGR gas is flowing and causing a higher EGR temperature reading.

Answer D is incorrect. Exhaust gas temperature will be much higher than what is displayed if the valve is overheated.

TASK B.21

14. An ECM has been diagnosed with a faulty injector driver. This could be caused by:

   A. Low circuit current
   B. High circuit resistance
   C. Low circuit resistance
   D. Low circuit voltage

   Answer A is incorrect. Low current will not damage the injector driver. High current through an injector driver will cause damage to the driver.

   Answer B is incorrect. High resistance will not damage the injector driver, since it will lower the current flow through the driver.

   Answer C is correct. A shorted fuel injector can cause low circuit resistance. The lower resistance can increase current flow, which will damage the injector driver.

   Answer D is incorrect. Low voltage will not damage the injector driver, since the current will be lower as well.

TASK A.14

15. A vehicle has been brought into the shop several times for a failed camshaft position (CMP) sensor. The root cause of this problem could be:

   A. Excessive cam thrust
   B. A slipped timing belt
   C. Improper engine oil
   D. Excessive timing belt tension

   Answer A is correct. Excessive cam thrust can damage a CMP sensor. If the thrust condition is not corrected, the problem will happen again.

   Answer B is incorrect. A slipped timing belt will not damage a CMP sensor, as this condition will not cause the cam to "walk."

   Answer C is incorrect. Improper engine oil will have no effect on the operation of the CMP sensor, but will affect the variable valve system if equipped.

   Answer D is incorrect. Excessive timing belt tension will not damage a CMP sensor, as this will not cause the cam to contact the CMP sensor.

TASK C.2

16. Spark knock can be caused by which of the following?

   A. Colder spark plugs than specified
   B. Hotter spark plugs than specified
   C. Smaller plug gap than specified
   D. Larger plug gap than specified

   Answer A is incorrect. Colder spark plugs than specified will not cause spark knock, because they will allow the combustion chamber to run cooler.

   Answer B is correct. The use of hotter spark plugs will hold more heat in the cylinder, which could lead to spark knock.

   Answer C is incorrect. A smaller spark plug gap than specified will not cause spark knock; it could cause higher emissions and low power output.

   Answer D is incorrect. A larger spark plug gap than specified will not cause spark knock; it could cause the engine to misfire and cause secondary ignition component damage.

17. Which of the following would LEAST LIKELY cause a high percentage IAC command by the ECM?

TASK B.8

    A. A dirty throttle plate

    B. Dirty IAC passages

    C. A faulty IAC valve

    D. A vacuum leak

Answer A is incorrect. A dirty throttle plate would cause the ECM to command a high IAC percentage to bring the idle speed up to spec.

Answer B is incorrect. Dirty IAC passages would restrict the amount of air allowed into the engine, so the IAC would have to open more to compensate for the lack of airflow.

Answer C is incorrect. If the IAC valve were faulty and could not move, the ECM would command it to open, but the IAC would not be able to respond to the command

Answer D is correct. A vacuum leak on a speed density fuel controlled vehicle will cause a higher-than-normal idle speed. This will cause the ECM to lower the commanded IAC percentage.

18. A failed catalytic converter would cause lower output of which exhaust gas?

TASK E.10

    A. NO_x

    B. CO_2

    C. HC

    D. O_2

Answer A is incorrect. NO_x emissions would increase if the catalytic converter were to fail.

Answer B is correct. Since the catalytic converter is not further burning the HC gases and combining CO and O_2, CO_2 levels would be lower than normal.

Answer C is incorrect. HC emissions would increase if the catalytic converter were to fail.

Answer D is incorrect. O_2 emissions would increase if the catalytic converter were to fail.

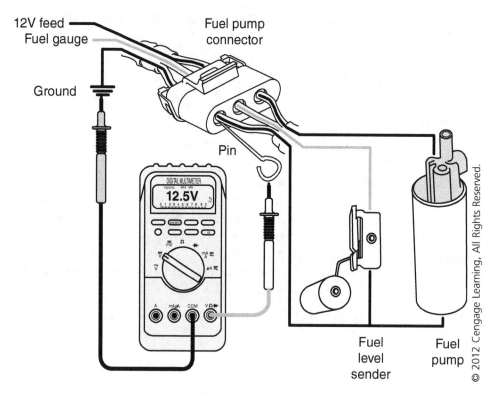

TASK D.13

19. A suspected faulty fuel pump was replaced and did not correct the concern. The meter above displays 12.5 V. Technician A says that this reading is normal when the fuel pump is not commanded on. Technician B says that this could indicate a faulty fuel pump if the pump is being commanded on. Who is correct?

   A. A only

   B. B only

   C. Both A and B

   D. Neither A nor B

Answer A is incorrect. When the fuel pump is not turned on, the reading should show zero volts.

Answer B is incorrect. This measurement cannot be used alone to determine if the fuel pump is good or bad. It will only determine the integrity of the connection.

Answer C is incorrect. Neither Technician is correct.

Answer D is correct. Neither Technician is correct. A reading of 12.5 V indicates that the ground side is open.

20. A vehicle starts and dies. Technician A says that a malfunction in the anti-theft system can cause these symptoms. Technician B says that a malfunctioning IAC valve can cause this condition. Who is correct?

TASK B.13

   A. A only
   B. B only
   C. Both A and B
   D. Neither A nor B

   Answer A is incorrect. Technician B is also correct.

   Answer B is incorrect. Technician A is also correct.

   Answer C is correct. A malfunction in the anti-theft system will cause the vehicle to start and then immediately die. A stuck closed IAC valve will also cause the vehicle to die immediately after starting.

   Answer D is incorrect. Both Technicians are correct.

21. A vehicle failed the initial idle exhaust emissions test, but passed the second-chance test. Technician A says that the vehicle was not preconditioned before the test. Technician B says that the catalytic converter may not have been hot enough. Who is correct?

TASK F.16

   A. A only
   B. B only
   C. Both A and B
   D. Neither A nor B

   Answer A is incorrect. Technician B is also correct.

   Answer B is incorrect. Technician A is also correct.

   Answer C is correct. Both Technicians are correct. A vehicle that was not properly preconditioned for an emissions test may have a cold catalytic converter. If the catalytic converter is not warm enough for operation, the vehicle will fail the emissions test.

   Answer D is incorrect. Both Technicians are correct.

22. A vehicle was brought into the shop as a no start. It was found that the distributor rotor had a hole in it. The most likely cause for this is:

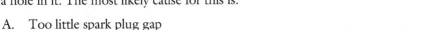

TASK C.9

   A. Too little spark plug gap
   B. A shorted ignition coil
   C. High coil wire resistance
   D. High secondary wire resistance

   Answer A is incorrect. Too little spark plug gap would decrease secondary resistance, which would not damage the rotor.

   Answer B is incorrect. A shorted ignition coil would not damage the distributor rotor. A short on the primary side would damage the coil driver or module, while a short on the secondary side would not produce enough voltage to burn a hole in the rotor.

   Answer C is incorrect. Since the coil wire is before the distributor rotor, high resistance would not damage the rotor; it would most likely damage the coil.

   Answer D is correct. High resistance in one of the secondary spark plug wires can cause the high voltage to burn through the distributor cap or rotor.

SCAN TOOL DATA

Engine Coolant Temperature Sensor (ECT)	Intake Air Temperature Sensor (IAT)	Manifold Absolute Pressure Sensor (MAP)	Throttle Position Sensor (TPS)
0.42 Volts	2.24 Volts	2.75 Volts	0.6 Volts
Engine Speed Sensor (RPM)	Heated Oxygen Sensor (HO$_2$S)	Vehicle Speed Sensor (VSS)	Battery Voltage (B+)
750 rpm	0.3–0.7 Volts	0 mph	14.2 Volts
Idle Air Control Valve (IAC)	Evaporative Emission Canister Solenoid (EVAP)	Torque Converter Clutch Solenoid (TCC)	EGR Valve Control Solenoid (EGR)
30 percent	OFF	OFF	0 percent
Malfunction Indicator Lamp (MIL)	Diagnostic Trouble Codes	Open/Closed Loop	Fuel Pump Relay (FP)
OFF	NONE	CLOSED	ON

Measured Ignition Timing °BTDC	Base Timing: 10°	Actual Timing: 11°

TASK B.20

23. Technician A says that a lean operating engine can cause the data that is shown in the above figure. Technician B says that the data shown can be caused by a plugged catalytic converter. Who is correct?

    A. A only

    B. B only

    C. Both A and B

    D. Neither A nor B

    Answer A is incorrect. There is no data shown, such as low O_2 voltage or high fuel trims, that suggests a lean operating engine.

    Answer B is correct. Only Technician B is correct. The high MAP sensor voltage could be caused by a plugged catalytic converter.

    Answer C is incorrect. Only Technician B is correct.

    Answer D is incorrect. Technician B is correct.

TASK A.8

24. A vehicle owner has installed a colder-than-specified thermostat in the vehicle. This will affect vehicle emissions by:

    A. Increasing HC emissions

    B. Increasing NO_x emissions

    C. Increasing CO_2 emissions

    D. Decreasing CO emissions

    Answer A is correct. A colder operating engine will not burn fuel as efficiently, and HCs will be able to leave the combustion chamber unburned.

    Answer B is incorrect. NO_x emissions are formed under high temperature operating conditions.

    Answer C is incorrect. A colder thermostat will cause the engine to run rich, which will cause CO_2 levels to decrease.

    Answer D is incorrect. A colder thermostat will cause the air/fuel ratio to go rich, which will cause CO emissions to increase.

25. Measured current draw on a fuel pump circuit is lower than normal. This could be caused by:

    A. A short to ground
    B. An open fuel pump wire
    C. An open relay coil
    D. A corroded connector

TASK B.15

Answer A is incorrect. A short to ground would increase current flow and possibly blow a fuse.

Answer B is incorrect. An open fuel pump wire would cause no current to flow in the circuit at all.

Answer C is incorrect. An open relay coil would cause no current to flow in the circuit at all.

Answer D is correct. A corroded connector would cause high resistance and lower the amount of current flowing through the circuit.

26. Positive fuel trim numbers could be caused by:

    A. Excessive fuel pressure
    B. A clogged fuel filter
    C. A leaking injector
    D. A leaking purge valve

TASK D.5

Answer A is incorrect. Excessive fuel pressure would cause a rich condition, which would cause fuel trim numbers to be negative.

Answer B is correct. A clogged fuel filter will cause fuel pressure to be lower than normal, which would cause a lean condition. The ECM would have to add fuel to achieve a 14.7:1 air/fuel ratio. This would cause fuel trim numbers to increase into positive numbers.

Answer C is incorrect. A leaking injector would cause a rich condition, which would cause fuel trim numbers to be negative.

Answer D is incorrect. A leaking purge valve would cause a rich condition, which would cause fuel trim numbers to be negative.

27. Technician A says that many states are not equipped to perform the dyno test on all-wheel drive vehicles. Technician B says that all-wheel drive vehicles are usually exempt from emissions testing because they cannot be dyno tested. Who is correct?

    A. A only
    B. B only
    C. Both A and B
    D. Neither A nor B

TASK F.16

Answer A is correct. Only Technician A is correct. Many emissions testing facilities are not equipped to perform the dyno test on all-wheel drive vehicles, so they are exempt from that portion of the test.

Answer B is incorrect. Even though all-wheel drive vehicles cannot be tested on the dyno, these vehicles will still be required to have either a plug-in test or a two speed idle test.

Answer C is incorrect. Only Technician A is correct.

Answer D is incorrect. Technician A is correct.

TASK E.8

28. The o-ring seal on the gas cap has many small cracks. Technician A says that these cracks can cause the MIL to illuminate and an EVAP DTC to set if the leak is larger than 0.020" (0.508 mm). Technician B says that the cracks can cause the vehicle to fail an I/M test. Who is correct?

    A. A only

    B. B only

    C. Both A and B

    D. Neither A nor B

    Answer A is incorrect. Technician B is also correct.

    Answer B is incorrect. Technician A is also correct.

    Answer C is correct. Both Technicians are correct. Any leak larger than 0.020" (0.508 mm) may cause the MIL to illuminate. A gas cap test is part of many I/M test procedures. If the cap is leaking, it will cause the vehicle to fail the test.

    Answer D is incorrect. Both Technicians are correct.

TASK B.7

29. A vehicle entered the shop with a DTC. The vehicle was diagnosed and repaired, and the DTC was cleared. To verify the repair, the technician should:

    A. Test drive the vehicle while monitoring scan tool data

    B. Refer to the DTC enabling criteria and test drive

    C. Perform a functional test of the engine control system

    D. Inform the customer to return for a follow-up inspection

    Answer A is incorrect. A simple test drive may not put the vehicle under the conditions required to run the test for the DTC.

    Answer B is correct. After the enabling criteria for the DTC is consulted, the vehicle should be test driven and operated under the conditions of the enabling criteria. If the DTC does not reset, the repair can be verified.

    Answer C is incorrect. A complete functional test of the engine controls system is not an efficient way to verify a repair and is also unnecessary.

    Answer D is incorrect. Asking the customer to return for a follow-up inspection is not a professional or acceptable way of verifying a repair.

TASK F.10

30. A vehicle fails a loaded I/M emissions test for high HC emissions. HC emissions were acceptable at idle, but increased under a load. This could be caused by:

    A. A leaking fuel injector

    B. A vacuum leak

    C. Excessive fuel pressure

    D. Excessive plug wire resistance

    Answer A is incorrect. A leaking fuel injector would cause high HC emissions at idle.

    Answer B is incorrect. A vacuum leak may cause a lean misfire at idle but would affect HC emissions less under a load.

    Answer C is incorrect. Excessive fuel pressure would cause high HC emissions at idle.

    Answer D is correct. At idle, the coil may be able to supply the required voltage of the cylinder through a spark plug wire with high resistance. Under a load, the cylinder's required voltage will increase, which may cause the spark to find an easier path to ground and cause a miss. When this occurs, HC levels will increase.

31. While a vehicle with an on-board refueling vapor recovery (ORVR) system is being refueled, fuel vapors are:

TASK E.4

   A. Vented to the air intake

   B. Stored in the fuel tank

   C. Stored in the EVAP canister

   D. Vented to the atmosphere

Answer A is incorrect. If vapors were vented to the intake, it would be possible for them to escape into the atmosphere.

Answer B is incorrect. There is not enough room left in the fuel tank for vapor storage after the tank is full.

Answer C is correct. Fuel vapors are stored in the EVAP canister so that they can be burned in the engine later.

Answer D is incorrect. Fuel vapors should never be vented to the atmosphere.

32. A lower-than-normal MAF sensor reading can be caused by all of the following EXCEPT:

TASK B.17

   A. A vacuum leak

   B. A restricted exhaust

   C. MAF contamination

   D. High fuel pressure

Answer A is incorrect. A vacuum leak will allow unmetered air into the engine, which will not be measured by the MAF sensor. This will cause it to read lower than normal.

Answer B is incorrect. A restricted exhaust will cause airflow through the engine to be less than normal. This smaller amount of air will cause MAF sensor readings to be low.

Answer C is incorrect. A contaminated MAF sensor will not be able to accurately measure the incoming air. Less than the actual airflow will be measured.

Answer D is correct. High fuel pressure will not affect the reading of the MAF sensor.

33. A customer is concerned that water is dripping out of the vehicle's tailpipe. Technician A says that this could be caused by a malfunctioning catalytic converter. Technician B says that the vehicle is running too lean. Who is correct?

TASK E.1

   A. A only

   B. B only

   C. Both A and B

   D. Neither A nor B

Answer A is incorrect. A malfunctioning catalytic converter will not cause water to form in the exhaust. Other emissions would be affected if the converter were malfunctioning.

Answer B is incorrect. Water coming out of the tailpipe is a normal condition and cannot be used to determine if the engine is running lean.

Answer C is incorrect. Neither Technician is correct.

Answer D is correct. Neither Technician is correct. Water coming from the tailpipe of a vehicle can be normal, especially when it is cold. Water forms in the exhaust when hydrogen from the burned fuel combines with oxygen in the air.

TASK B.8

34. To best find out what conditions were present when a DTC set in the vehicle's ECM, the technician should:

    A.   Refer to freeze frame data
    B.   Refer to the DTC enabling criteria
    C.   Question the customer
    D.   Clear the code and attempt to reset it

    Answer A is correct. When trying to determine what conditions were present when a DTC set, the technician should consult the freeze frame data for that code.

    Answer B is incorrect. The enabling criteria will not report the exact vehicle conditions present when the DTC set.

    Answer C is incorrect. While it is good to question the customer about the concern, the customer is not likely to know the sensor data or engine conditions that were present when the code was set.

    Answer D is incorrect. A code should not be cleared in an attempt to find the conditions for setting the code. This will clear the freeze frame data that shows the conditions for setting the DTC.

TASK F.16

35. A vehicle has failed an I/M test for high CO emissions. This could be caused by:

    A.   Overfilling the fuel tank
    B.   Low fuel system pressure
    C.   An intake manifold vacuum leak
    D.   An inoperative EGR valve

    Answer A is correct. Overfilling the fuel tank can saturate the EVAP charcoal canister with fuel. This excessive fuel can cause the vehicle to operate rich, which would increase CO emissions.

    Answer B is incorrect. Low fuel pressure would not cause increased CO emissions, since it would cause a lean condition.

    Answer C is incorrect. An intake manifold vacuum leak would cause a lean condition and higher than normal O_2 exhaust emissions.

    Answer D is incorrect. An inoperative EGR valve would only affect NO_x emissions.

TASK A.9

36. All of the following could increase HC exhaust emissions EXCEPT:

    A.   A vacuum leak
    B.   Stuck open thermostat
    C.   Burned exhaust valve
    D.   Clogged EGR passage

    Answer A is incorrect. A large vacuum leak could lead to a lean misfire, which could increase HC levels in the exhaust.

    Answer B is incorrect. An engine does not burn fuel as efficiently when cold. A stuck open thermostat would cause HC levels to increase.

    Answer C is incorrect. A burned exhaust valve would allow unburned HCs to enter the exhaust stream.

    Answer D is correct. Insufficient EGR flow would cause excessive NO_x emissions. HCs would not be affected.

37. What could cause a vehicle to fail the two-speed idle test for excessive CO emissions?

TASK F.9

    A. A leaking vacuum controlled fuel pressure regulator

    B. A leaking PCV system vacuum hose

    C. A restricted fuel filter

    D. A restricted catalytic converter

Answer A is correct. A leaking fuel pressure regulator could cause increased CO emissions due to the rich condition that it would create.

Answer B is incorrect. A leaking PCV vacuum hose would cause a lean condition, which would increase O_2 levels instead of CO levels.

Answer C is incorrect. A restricted fuel filter would cause a lean condition.

Answer D is incorrect. A restricted catalytic converter would cause low engine power, but it would not cause increased CO emissions.

38. The long-term fuel trim obtained from a vehicle's scan tool data is -20 percent. This data indicates:

TASK D.5

    A. The ECM is correcting for a lean condition

    B. The ECM is correcting for a rich condition

    C. A vacuum leak could be present

    D. The ECM is adding more fuel than normal

Answer A is incorrect. Negative fuel trim numbers indicate that a rich condition is present.

Answer B is correct. Negative fuel trim data indicates that the ECM is taking fuel away from the engine due to a rich condition.

Answer C is incorrect. A vacuum leak would cause fuel trim numbers to increase.

Answer D is incorrect. When fuel trim numbers are negative, the ECM is taking fuel away from the engine to compensate for a lean air/fuel ratio.

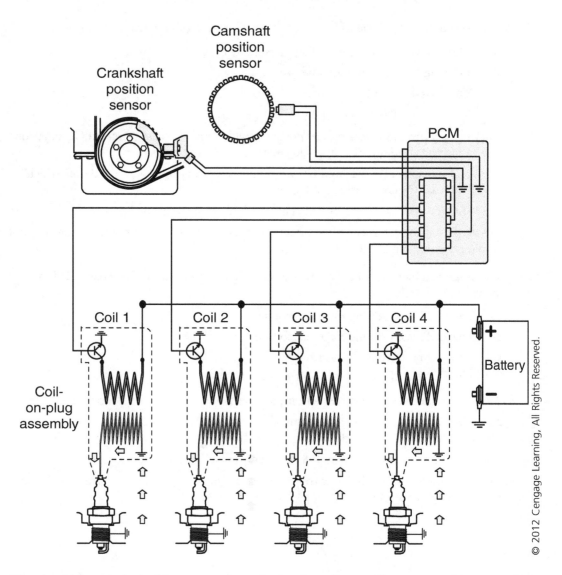

TASK C.8

39. A vehicle with the ignition system shown in the above figure will not start and is not sparking on any cylinder. All of the following could cause this EXCEPT:

   A. A faulty cam sensor
   B. A faulty coil
   C. A faulty crank sensor
   D. A faulty ECM

Answer A is incorrect. The cam sensor is used to determine coil timing. If the cam sensor were faulty, the ECM would not know which cylinder to send spark to.

Answer B is correct. A faulty coil would only cause one cylinder to misfire.

Answer C is incorrect. The crank sensor signal is needed for spark timing.

Answer D is incorrect. A faulty ECM could cause the coils not to spark.

40. A DTC P0401 – EGR insufficient flow has set on the composite vehicle. The EGR solenoid winding resistance is 10 KΩ. This reading indicates:

TASK E.9

A. A normal resistance reading

B. A shorted EGR solenoid

C. An open EGR solenoid

D. Excessive EGR solenoid resistance

Answer A is incorrect. The reading is higher than the provided specification of 12 Ω and is not normal.

Answer B is incorrect. A shorted solenoid winding would have a resistance of less than the 12 Ω specification.

Answer C is incorrect. An open solenoid winding would have no resistance at all and would display O.L. on the meter.

Answer D is correct. The resistance specification for the EGR solenoid is 12 ± 2 Ω. The reading shown is 10,000 Ω.

41. The composite vehicle's VVT system has failed, and the ECM cannot make any camshaft timing adjustments. Technician A says that no oil pressure to the actuator control solenoid can cause this condition. Technician B says that the cams will default to the fully advanced position when the VVT system fails. Who is correct?

TASK A.4

A. A only

B. B only

C. Both A and B

D. Neither A nor B

Answer A is correct. Only Technician A is correct. Since oil pressure advances the camshafts, a lack of oil pressure will not allow the cam timing to be changed.

Answer B is incorrect. When the VVT system fails to operate, the camshafts will default to the fully retarded position.

Answer C is incorrect. Only Technician A is correct.

Answer D is incorrect. Technician A is correct.

42. The composite vehicle's throttle is unresponsive. The RPM is at 1400, and the throttle is fixed at 15 percent. This could be caused by:

TASK D.9

A. One faulty accelerator pedal position (APP) sensor

B. One faulty TPS

C. A malfunctioning throttle actuator motor

D. A throttle actuator that needs to be relearned

Answer A is incorrect. One faulty APP sensor will cause the system to limit the maximum throttle opening to 35 percent.

Answer B is incorrect. One faulty TPS will cause the system to limit the maximum throttle opening to 35 percent.

Answer C is correct. A faulty throttle actuator motor will cause the throttle to default to the 15 percent open position and be unresponsive.

Answer D is incorrect. The composite vehicle does not have a relearn procedure for the throttle actuator.

TASK E.8

43. A failure in the PCV system could cause all of the following conditions EXCEPT:

    A. Excessive oil leaks

    B. Rough or unstable idle

    C. Increased fuel economy

    D. DTCs to be set in the ECM

Answer A is incorrect. A malfunctioning PCV system could cause increased crankcase pressure, which would cause oil leaks.

Answer B is incorrect. A leaking PCV vacuum hose could cause the vehicle to have a rough or unstable idle speed.

Answer C is correct. A malfunction in the PCV system will not cause increased fuel economy.

Answer D is incorrect. A leaking PCV vacuum hose could cause a DTC to set in the ECM.

SCAN TOOL DATA

Engine Coolant Temperature Sensor (ECT) 0.35 Volts	Intake Air Temperature Sensor (IAT) 4.50 Volts	MAP Sensor 4.5 V MAF Sensor 0.2 V	Throttle Position Sensor (TPS) 0.6 Volts
Engine Speed Sensor (RPM) 0 rpm	Heated O$_2$ HO$_2$S Upstream 0 V Downstream 0 V	Vehicle Speed Sensor (VSS) 0 mph	Battery Voltage (B+) 12.6 Volts
	Evaporative Emission Canister Solenoid (EVAP) OFF	Torque Converter Clutch Solenoid (TCC) OFF	EGR Valve Control Solenoid (EGR) 0 percent
Malfunction Indicator Lamp (MIL) ON	Diagnostic Trouble Codes NONE	Open/Closed Loop OPEN	Fuel Pump Relay (FP) OFF
Fuel Level Sensor 3.5 V	Fuel Tank Pressure Sensor 2.5 V	Transmission Fluid Temperature Sensor 3.5 V	Transmission Turbine Shaft Speed Sensor 0 mph
Transmission Range Switch PARK	Trans Pressure Control Solenoid 0	Transmission Shift Solenoid 1 ON	Transmission Shift Solenoid 2 OFF

Measured Ignition Timing °BTDC   Base Timing: 10°   Actual Timing: 20°

TASK B.8

44. The composite vehicle has entered the shop with a poor economy concern. Using the KOEO scan tool data shown in the above figure, determine the cause of the fuel economy concern.

    A. Faulty IAT sensor

    B. Faulty ECT sensor

    C. Faulty MAP sensor

    D. Faulty MAF sensor

Answer A is correct. The IAT sensor is reading a voltage that corresponds with −20° F (−29°C). This will cause more fuel to be added unnecessarily because of the colder temperature.

Answer B is incorrect. The ECT sensor is showing a warmed-up vehicle. Even if this were an incorrect reading, it would not cause a poor fuel economy concern.

Answer C is incorrect. The MAP sensor is reading the correct KOEO voltage. Low MAP sensor voltages with the engine off would indicate a malfunction.

Answer D is incorrect. The MAF sensor is reading the correct KOEO voltage. Any airflow indicated with the engine off should be cause for alarm.

45. The composite vehicle will not start, and there is no signal from the crank sensor. Technician A says that the voltage to the sensor should be measured. Technician B says that the resistance of the sensor should be measured. Who is correct?

TASK C.5

   A.   A only
   B.   B only
   C.   Both A and B
   D.   Neither A nor B

   Answer A is incorrect. The crank sensor generates an AC voltage and does not require an input voltage.

   Answer B is correct. Only Technician B is correct. The resistance of the crank sensor should be measured if it is not producing an output signal.

   Answer C is incorrect. Only Technician B is correct.

   Answer D is incorrect. Technician B is correct.

46. The composite vehicle will not start, and the scan tool will not communicate. Technician A says that an unplugged instrument panel cluster will cause the scan tool to not communicate. Technician B says that a data bus that is shorted to ground will cause this problem. Who is correct?

TASK B.12

   A.   A only
   B.   B only
   C.   Both A and B
   D.   Neither A nor B

   Answer A is incorrect. An unplugged instrument panel cluster will cause a network fault and DTC, but it will not cause a no-start, no-communication issue.

   Answer B is incorrect. A data bus that is shorted to ground will cause a no-communication condition, but the vehicle will still start.

   Answer C is incorrect. Neither Technician is correct.

   Answer D is correct. Neither Technician is correct. Although the data bus for the composite vehicle is a non-fault tolerant network, a failure in network communications will not prevent the ECM from providing ignition and fuel control.

47. The composite vehicle has a high idle. The cause of this condition could be:

TASK D.10

   A.   A sticking throttle cable
   B.   ECM terminal 172 shorted to power
   C.   ECM terminal 172 shorted to ground
   D.   No voltage to the A/C pressure switch

   Answer A is incorrect. The composite vehicle does not use a throttle cable.

   Answer B is correct. When the ECM detects power on terminal 172, the idle speed is raised to prevent stalling due to the extra load from the A/C compressor.

   Answer C is incorrect. If ECM terminal 172 were shorted to ground, fuse #56 would blow when the A/C request switch closed and many other symptoms would be present.

   Answer D is incorrect. No voltage to the A/C request switch would not allow the A/C to function at all.

TASK B.13

48. The composite vehicle has been towed into the shop with an immobilizer system failure. This failure will:

   A.   Not allow engine cranking
   B.   Turn the security indicator on constantly
   C.   Cause the engine to start and stall
   D.   Not allow scan tool communication

   Answer A is incorrect. The immobilizer system will not inhibit engine cranking.

   Answer B is incorrect. When a failure occurs in the immobilizer system, the security indicator will flash instead of remaining on constantly.

   Answer C is correct. When a failure has occurred in the immobilizer system, the ECM will shut the fuel injectors off after two seconds of running.

   Answer D is incorrect. A failure in the immobilizer will not affect scan tool communication.

TASK E.5

49. The composite vehicle has entered the shop with a DTC P0440 – EVAP system leak. Before injecting smoke to look for the leak, the technician should:

   A.   Command the purge solenoid closed
   B.   Block off the EVAP canister
   C.   Install a non-vented fuel cap
   D.   Command the vent solenoid closed

   Answer A is incorrect. The purge solenoid is normally closed; therefore, it does not have to be commanded closed.

   Answer B is incorrect. If the EVAP canister is blocked off, the leak may not be found.

   Answer C is incorrect. The smoke machine will not raise pressure enough to open the vacuum valve in the cap unless it is faulty.

   Answer D is correct. When diagnosing an EVAP system leak, the vent solenoid should be commanded shut to seal the system.

TASK F.14

50. The catalytic converter readiness monitor for the composite vehicle displays "Not Complete." This reading can be caused by:

   A.   A failed catalytic converter monitor
   B.   A failed oxygen sensor monitor
   C.   Miscommunication of the scan tool
   D.   A faulty ECM

   Answer A is incorrect. A message of "Not Complete" does not indicate a failed monitor; it only indicates that the monitor has not yet run.

   Answer B is correct. Before the catalytic converter monitor can be run, the O_2 sensor monitors must run and pass, since they are used to test the catalytic converter.

   Answer C is incorrect. If the scan tool displays any data at all, there cannot be a miscommunication of data.

   Answer D is incorrect. It is not likely that a faulty ECM will cause a monitor not to run.

PREPARATION EXAM 4—ANSWER KEY

1. C	**21.** C	**41.** D
2. C	**22.** D	**42.** A
3. A	**23.** B	**43.** C
4. B	**24.** A	**44.** B
5. C	**25.** D	**45.** D
6. B	**26.** B	**46.** C
7. D	**27.** C	**47.** C
8. B	**28.** A	**48.** D
9. A	**29.** A	**49.** D
10. D	**30.** D	**50.** D
11. B	**31.** C	
12. A	**32.** B	
13. D	**33.** A	
14. C	**34.** A	
15. D	**35.** D	
16. A	**36.** A	
17. C	**37.** C	
18. B	**38.** D	
19. C	**39.** B	
20. A	**40.** B	

PREPARATION EXAM 4—EXPLANATIONS

1. Technician A recommends testing the battery before reprogramming the electronic control module (ECM). Technician B recommends disabling the daytime running lamps while reprogramming. Who is correct?

    TASK B.6

    A. A only

    B. B only

    C. Both A and B

    D. Neither A nor B

    Answer A is incorrect. Technician B is also correct.

    Answer B is incorrect. Technician A is also correct.

    Answer C is correct. Both Technicians are correct. The battery should be fully charged and able to power the vehicle while the ECM is being programmed, so it is recommended that it be tested. It is also advisable to turn off all vehicle loads, especially DRLs or automatic headlamps that may illuminate in the shop.

    Answer D is incorrect. Both Technicians are correct.

TASK A.2

2. Technician A says that installing a high-flow exhaust system will affect the operation of the exhaust gas recirculation (EGR) system. Technician B says an exhaust system of this type may not be legal. Who is correct?

A. A only

B. B only

C. Both A and B

D. Neither A nor B

Answer A is incorrect. Technician B is also correct.

Answer B is incorrect. Technician A is also correct.

Answer C is correct. Both Technicians are correct. A high-flow exhaust system may not provide enough backpressure for the EGR system to operate properly, and this type of exhaust system may not be approved for street use.

Answer D is incorrect. Both Technicians are correct.

TASK D.10

3. After sitting overnight, a vehicle starts, stalls, and dies unless the throttle is pressed. After warming up, the vehicle will idle, but RPM is still low. What could cause this problem?

A. A dirty idle air control (IAC) valve

B. An intake manifold vacuum leak

C. A dirty air filter

D. A leaking fuel pump check valve

Answer A is correct. A dirty IAC valve would cause the above symptoms. It is not possible to clean the entire valve, so it should be replaced.

Answer B is incorrect. A vacuum leak is likely to cause a rough idle at all engine temperatures (on MAF vehicles) or a high idle (on MAP vehicles).

Answer C is incorrect. A dirty air filter may cause low engine power, but it is not likely to stop the engine from running.

Answer D is incorrect. A leaking fuel pump check valve will lead to a long crank before starting concern.

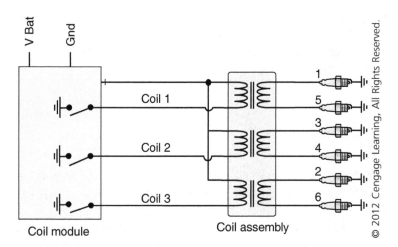

4. Technician A says that a fault in the ground circuit of one of the pictured coils will cause a miss on one cylinder. Technician B says that an open in the power wire will cause the vehicle not to start. Who is correct?

TASK C.8

   A.   A only

   B.   B only

   C.   Both A and B

   D.   Neither A nor B

Answer A is incorrect. One coil of this distributorless electronic ignition (EI) system controls the spark of two cylinders. If one coil becomes inoperative, two cylinders will be without spark.

Answer B is correct. Only Technician B is correct. Since all of the coils receive power from the same wire, an open in this wire would cause all three coils to be inoperative.

Answer C is correct. Only Technician B is correct.

Answer D is incorrect. Technician B is correct.

5. A vehicle that failed the I/M test for excessive NO_x emissions will likely have to be tested under what vehicle condition to duplicate the concern?

TASK F.1

   A.   Idle

   B.   Heavy acceleration

   C.   Light acceleration

   D.   Deceleration

Answer A is incorrect. Very little NO_x forms during idle.

Answer B is incorrect. Very little NO_x forms during heavy acceleration.

Answer C is correct. The most NO_x formation occurs under light acceleration.

Answer D is incorrect. Very little NO_x forms during deceleration.

TASK E.9

6. A vehicle enters the shop with a DTC P0440 – EVAP system small leak detected. Technician A says that shop air can be used to pressurize the system as long as it is regulated down to 0.5 psi (3.4 kPa). Technician B says that spraying soapy water around a pressurized system will allow the technician to locate the leak. Who is correct?

 A. A only

 B. B only

 C. Both A and B

 D. Neither A nor B

Answer A is incorrect. An EVAP system should never be pressurized using shop air, as this could cause a fire. Nitrogen should be the only gas used to pressurize the system, because it is inert and not flammable.

Answer B is correct. Only Technician B is correct. While pressurizing the system with nitrogen, the technician can spray various areas with soapy water to locate the leak.

Answer C is incorrect. Only Technician B is correct.

Answer D is incorrect. Technician B is correct.

TASK B.16

7. A faulty IAT sensor is suspected of causing a poor fuel economy complaint. Technician A says that the sensor voltage should steadily increase as the sensor heats up. Technician B says that sensor resistance should steadily increase as the sensor heats up. Who is correct?

 A. A only

 B. B only

 C. Both A and B

 D. Neither A nor B

Answer A is incorrect. The sensor voltage should steadily decrease as it heats up.

Answer B is incorrect. The sensor resistance should steadily decrease as it heats up.

Answer C is incorrect. Neither Technician is correct.

Answer D is correct. Neither Technician is correct. Most temperature sensors used on the engine are negative temperature coefficient type thermistors. As they are heated, their resistance goes down.

TASK A.11

8. A poor-quality air filter has been installed in a vehicle, and the MAF sensor is now contaminated. Which code will this condition most likely set?

 A. P0174 – system rich

 B. P0171 – system lean

 C. P0101 – MAF sensor performance

 D. P0131 – O_2 sensor low voltage

Answer A is incorrect. A contaminated MAF sensor will cause a lean condition.

Answer B is correct. A contaminated MAF sensor will measure less airflow than is present. Fuel will be delivered based on the lower amount of airflow, and a lean condition will exist.

Answer C is incorrect. Since the MAF sensor is functioning and changing according to the throttle position sensor (TPS) and RPM, the ECM will consider the MAF sensor reading to be correct.

Answer D is incorrect. The O_2 sensor will still be able to switch when the ECM adds fuel to compensate for the lean condition, so this code will not set.

9. Technician A says that a vacuum leak will cause lower-than-normal MAF sensor voltage at idle. Technician B says that a vacuum leak will cause lower-than-normal MAP sensor voltage at idle. Who is correct?

TASK B.8

   A. A only

   B. B only

   C. Both A and B

   D. Neither A nor B

Answer A is correct. Only Technician A is correct. A vacuum leak will allow unmetered air into the engine, which will cause the MAF sensor voltage to be lower than normal at idle.

Answer B is incorrect. A vacuum leak will increase intake manifold pressure, which will raise MAP sensor voltage.

Answer C is incorrect. Only Technician A is correct.

Answer D is incorrect. Technician A is correct.

10. A long-term fuel trim of 20 percent is indicated on the scan tool. Technician A says that this indicates a rich condition. Technician B says that this could be the result of excessive fuel pressure caused by an unapproved aftermarket fuel pressure regulator. Who is correct?

TASK D.2

   A. A only

   B. B only

   C. Both A and B

   D. Neither A nor B

Answer A is incorrect. Positive fuel trim numbers indicate a lean condition.

Answer B is incorrect. Excessive fuel pressure will cause a rich condition.

Answer C is incorrect. Neither Technician is correct.

Answer D is correct. Neither Technician is correct. Increased long term fuel trim numbers are an indication of a lean air/fuel ratio such as that caused by a vacuum leak. The ECM is adding fuel in an attempt to obtain a 14.7:1 air/fuel ratio.

11. A cracked spark plug will be displayed on the secondary voltage pattern as:

TASK C.10

   A. A higher firing line

   B. A lower firing line

   C. A shorter dwell period

   D. A longer dwell period

Answer A is incorrect. High firing lines are associated with high resistance. A cracked spark plug is considered low resistance.

Answer B is correct. If a spark plug becomes cracked, the voltage will find ground through the crack. This will require less voltage than usual, which will be displayed as a low firing line.

Answer C is incorrect. A cracked spark plug will not affect the dwell period, since dwell is controlled by the primary circuit and the plug is part of the secondary circuit.

Answer D is incorrect. A cracked spark plug will not affect the dwell period. Dwell is controlled and commanded by the ECM based on the engine's operating conditions.

TASK B.19

12. To test the operation of an O_2 sensor, a vacuum line is removed. The voltage of the sensor should:

    A. Decrease

    B. Increase

    C. Fluctuate rapidly

    D. Show no change

    Answer A is correct. The lean condition created by the vacuum leak should be displayed by a decrease in O_2 sensor voltage.

    Answer B is incorrect. An increase in O_2 sensor voltage indicates a rich condition.

    Answer C is incorrect. The lean condition would cause the sensor voltage to drop. Rapid fluctuation should be considered normal.

    Answer D is incorrect. O_2 sensor voltage should drop when a lean condition exists. An unresponsive sensor could indicate a fault with the sensor or the wiring.

TASK E.12

13. A low voltage condition displayed by the O_2 sensor will LEAST LIKELY be caused by:

    A. Low fuel pressure

    B. A vacuum leak

    C. The EVAP purge valve stuck open

    D. A cylinder misfire

    Answer A is incorrect. Low fuel pressure would cause a lean condition, which will cause O_2 sensor voltage to be low.

    Answer B is incorrect. A vacuum leak would cause a lean condition, which would cause the O_2 sensor to display low voltage.

    Answer C is incorrect. An EVAP purge valve that is stuck open will cause a rich condition and high O_2 sensor voltages.

    Answer D is correct. A cylinder misfire would cause the O_2 sensor to read lean and display low voltage, since the oxygen in that cylinder is not being burned.

TASK A.11

14. The results of a cracked exhaust manifold would be:

    A. Higher engine RPM

    B. Higher O_2 sensor voltage

    C. Lower O_2 sensor voltage

    D. Lower engine RPM

    Answer A is incorrect. A cracked exhaust manifold would not affect engine speed, because it is an unrelated component.

    Answer B is incorrect. High O_2 sensor readings occur when there is a lack of oxygen in the exhaust. An exhaust leak would allow excess oxygen into the exhaust.

    Answer C is correct. A cracked exhaust manifold would allow oxygen into the exhaust stream, and this would be shown as a low O_2 sensor voltage.

    Answer D is incorrect. A cracked exhaust manifold would not affect engine speed, as it has no relation to engine speed.

15. Technician A says that when CO levels are high the technician will have to diagnose a lean condition. Technician B says that when O_2 levels are high the technician will have to diagnose a rich condition. Who is correct?

TASK F.7

   A. A only

   B. B only

   C. Both A and B

   D. Neither A nor B

   Answer A is incorrect. When CO levels are high, a rich condition exists.

   Answer B is incorrect. When O_2 levels are high, a lean condition exists.

   Answer C is incorrect. Neither Technician is correct.

   Answer D is correct. Neither Technician is correct. High CO levels indicate that the engine is operating rich and receiving either too much fuel or too little air. High O_2 levels indicate that the engine is operating too lean and either receiving too little fuel or too much air.

16. The ignition rotor of a distributor ignition (DI) system has failed several times and created a no-start condition. The root cause of this failure could be which of the following?

TASK C.14

   A. High plug wire resistance

   B. A cracked spark plug

   C. High coil wire resistance

   D. A cracked distributor cap

   Answer A is correct. High spark plug resistance will cause the high secondary voltage to seek a shorter path to ground. This path can be through the rotor to the distributor shaft. If the component that caused the rotor to fail is not found and corrected, the rotor will fail repeatedly.

   Answer B is incorrect. A cracked spark plug will cause the engine to misfire, but it will not damage the rotor.

   Answer C is incorrect. High coil wire resistance will damage the ignition coil but not the rotor.

   Answer D is incorrect. A cracked distributor cap may cause the engine not to start, but it will not damage the rotor.

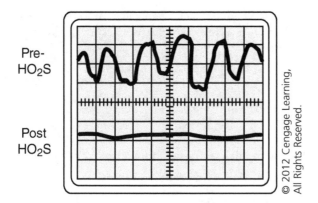

Pre-HO₂S

Post HO₂S

TASK B.17

17. What do the pictured O_2 sensor patterns indicate?

  A. A malfunctioning post HO_2S
  B. A malfunctioning pre-HO_2S
  C. A functioning catalytic converter
  D. A malfunctioning catalytic converter

  Answer A is incorrect. The post HO_2S is functioning normally and is indicating that the catalytic converter is also functioning.

  Answer B is incorrect. The pre-HO_2S is functioning normally and showing the constant switch from rich to lean.

  Answer C is correct. The steady voltage signal from the post HO_2S indicates that the catalytic converter is functioning properly.

  Answer D is incorrect. Should the catalytic converter fail, the post HO_2S will display a voltage similar to the pre-HO_2S.

TASK F.14

18. Technician A says that if all of the vehicle monitors have run and the MIL is not illuminated, all of the monitors have passed. Technician B says that if the MIL is illuminated, one or more of the monitors have failed. Who is correct?

  A. A only
  B. B only
  C. Both A and B
  D. Neither A nor B

  Answer A is incorrect. Some monitors require two consecutive failures before the MIL will illuminate. If the monitor runs and fails once, the MIL will not illuminate until the monitor runs and fails again.

  Answer B is correct. Only Technician B is correct. An illuminated MIL indicates that one or more monitors have failed.

  Answer C is incorrect. Only Technician B is correct.

  Answer D is incorrect. Technician B is correct.

19. A vehicle that continually goes into open loop at idle could be the result of:

TASK D.13

   A. A continuously lean condition
   B. A continuously rich condition
   C. A faulty O_2 sensor heater
   D. Poor fuel quality

   Answer A is incorrect. A lean condition will not affect loop status but will cause fuel trim numbers to increase.

   Answer B is incorrect. A rich condition will not affect loop status but will cause fuel trim numbers to decrease.

   Answer C is correct. An open O_2 sensor heater can allow the sensor to cool off at idle. This will result in the fuel system going into open loop, since the sensor is not hot enough for operation.

   Answer D is incorrect. Fuel quality will not affect loop status, but it may affect fuel trims.

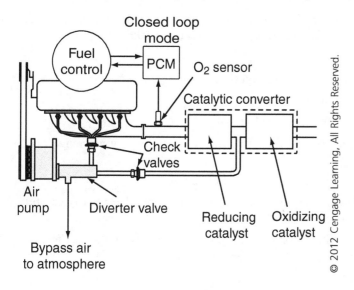

20. Under the conditions shown in the above image, where will the air output from the air pump go?

TASK E.5

   A. Downstream
   B. Upstream
   C. To the air cleaner
   D. No air will be pumped

   Answer A is correct. Under closed loop conditions, air from the air pump will be directed downstream to aid the catalytic converter in oxidation.

   Answer B is incorrect. If air were directed upstream during closed loop operation, the O_2 sensor would give a leaner signal than what is actually present.

   Answer C is incorrect. Air is directed to the air cleaner on deceleration.

   Answer D is incorrect. Air will always be pumped when using a belt-driven air pump.

TASK B.11

21. A vehicle starts and dies. Technician A says that a faulty IAC valve can cause this condition. Technician B says that a failure in the immobilizer system can cause this condition. Who is correct?

    A.   A only

    B.   B only

    C.   Both A and B

    D.   Neither A nor B

Answer A is incorrect. Technician B is also correct.

Answer B is incorrect. Technician A is also correct.

Answer C is correct. Both Technicians are correct. An IAC valve that is not functioning can cause the vehicle to start and die due to a lack of airflow. A fault in the immobilizer system will also cause the fuel injectors to shut off after two seconds.

Answer D is incorrect. Both Technicians are correct.

TASK A.15

22. After installing a new intake manifold gasket, which of the following methods is the safest and most effective way to locate an intake manifold vacuum leak?

    A.   Spray the intake with throttle body cleaner

    B.   Spray the intake with a soapy water solution

    C.   Blow compressed shop air into the intake

    D.   Connect a smoke machine to a vacuum line

Answer A is incorrect. Spraying any flammable chemical around a running engine could pose the threat of a fire.

Answer B is incorrect. Soapy water will not locate the leak.

Answer C is incorrect. Using compressed air in the intake could damage gaskets and will not give an accurate location of the leak.

Answer D is correct. Injecting smoke into the intake is the safest and most effective method of locating a vacuum leak.

SCAN TOOL DATA

Engine Coolant Temperature (ECT) Sensor 0.51 volts	Intake Air Temperature (IAT) Sensor 2.4 volts	Manifold Absolute Pressure (MAP) Sensor 1.8 volts	Throttle Position Sensor (TPS) 40%
Engine Speed Sensor (rpm) 1,700 rpm	Air Fuel Ratio Sensor AFRS 1/1 - 3.7 volts AFRS 2/1 - 3.7 volts	Vehicle Speed Sensor (VSS) 55 mph	Battery Voltage (B+) 14.3 volts
Fuel Pressure (FP) - 1.4 volts	Evaporative Emission Canister Solenoid (EVAP) ON	Torque Converter Clutch (TCC) Solenoid ON	EGR Valve Control (EGR) 30 percent
Malfunction Indicator Lamp (MIL) ON	Diagnostic Trouble Codes P0171*, P0174**, P0087***	Open/Closed Loop CLOSED	Fuel Pump 100%

*P0171 - Bank 1 System Lean
**P0174 - Bank 2 System Lean
***P0087 - Fuel Pressusre (FP) Sensor Low

23. The composite vehicle has low power under most driving conditions. Using the scan tool data shown, which of these would cause this condition?

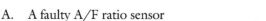

TASK B.8

   A. A faulty A/F ratio sensor

   B. A faulty fuel injector

   C. A weak fuel pump

   D. A faulty Fuel Pressure (FP) sensor

Answer A is incorrect. Since there are no codes for the A/F ratio sensors, and both banks indicate a lean condition, it is not likely that the A/F ratio sensors are faulty.

Answer B is incorrect. One faulty fuel injector would only impact one sensor. The fault that is present in this vehicle is affecting the entire engine.

Answer C is correct. A weak fuel pump would cause a low power condition under many driving conditions. Low fuel pressure is indicated by the fuel pressure (FP) sensor voltage. The lean codes and low fuel pressure code also point toward a faulty fuel pump.

Answer D is incorrect. Since DTCs and A/F ratio sensor voltages point toward a lean condition, a faulty fuel pressure (FP) sensor should not be suspected.

24. A failed catalytic converter would likely cause which of the following exhaust emissions to be higher at idle?

A. HC and CO

B. HC and NO$_x$

C. CO and CO$_2$

D. CO$_2$ and O$_2$

Answer A is correct. A failed catalytic converter will not be able to oxidize HC and CO to lower these emissions.

Answer B is incorrect. While a failed catalytic converter will not be able to oxidize HC, NO$_x$ levels will not be high at idle and will not increase. High NO$_x$ levels typically occur at higher RPMs instead of idle.

Answer C is incorrect. While a failed catalytic converter will not be able to oxidize CO, CO$_2$ levels are at their highest when everything is working. A failed converter will lower CO$_2$ levels.

Answer D is incorrect. A failed catalytic converter will not cause CO$_2$ or O$_2$ levels to increase. Any malfunctions that affect emissions will cause CO$_2$ levels to drop. O$_2$ levels are typically controlled by the air/fuel ratio of the vehicle.

25. What will the ECM do when the O$_2$ sensor shows an overall lean reading?

A. Decrease fuel trim

B. Increase idle speed

C. Decrease fuel delivery

D. Increase fuel trim

Answer A is incorrect. Fuel trim will be decreased when a rich condition exists, indicating that the engine is taking fuel away from the engine to achieve the proper air/fuel ratio.

Answer B is incorrect. The ECM will not command an idle speed change when a lean condition exists, because it will not correct the concern.

Answer C is incorrect. Fuel delivery will be decreased when a rich condition exists.

Answer D is correct. Fuel trim will be increased when a lean condition exists, indicating that the engine is adding fuel to the engine to achieve the proper air/fuel ratio.

26. An engine coolant temperature sensor is being tested for operation. While the resistance is being tested as the engine is warming up, the value should:

A. Steadily increase

B. Steadily decrease

C. Fluctuate rapidly

D. Read O.L.

Answer A is incorrect. The resistance of the sensor should go down as it is heated, because it is a negative temperature coefficient thermistor.

Answer B is correct. As the temperature goes up, the resistance of the sensor should go down.

Answer C is incorrect. The resistance of the sensor should not fluctuate; this would indicate a faulty sensor.

Answer D is incorrect. A reading of O.L. would indicate that the sensor is open.

27. A vehicle has a rough idle and excessive HC emissions at idle, but it runs smoothly and has acceptable HC emissions at 2000 rpm. Which of the following could cause this condition?

TASK F.8

   A. High spark plug wire resistance

   B. A clogged fuel filter

   C. A leaking intake runner gasket

   D. A fouled spark plug

Answer A is incorrect. High spark plug wire resistance would cause HCs to increase under a load.

Answer B is incorrect. A clogged fuel filter may cause a lean misfire at 2000 rpm, but no excessive HC emissions would be created at idle.

Answer C is correct. A leaking intake runner gasket can cause a lean misfire on one cylinder at idle. At higher RPMs the added fuel will allow a burnable air/fuel ratio.

Answer D is incorrect. A fouled spark plug would have elevated HC emissions under any engine condition.

28. The operation of the purge valve is being tested. With the purge valve open KOER, O_2 sensor voltage should:

TASK B.16

   A. Increase

   B. Decrease

   C. Remain the same

   D. Fluctuate rapidly

Answer A is correct. Opening the purge valve will cause a rich condition, which will cause the O_2 sensor voltage to increase.

Answer B is incorrect. Low O_2 sensor voltage is created by a lean condition. Opening the purge valve will cause a rich condition.

Answer C is incorrect. O_2 sensor voltage will not remain the same, as the air/fuel mixture will become richer with the purge valve open.

Answer D is incorrect. The air/fuel mixture will remain rich as long as the purge valve is open. The O_2 sensor voltage will remain high as long as the mixture remains rich.

29. A vehicle is brought to the shop for an illuminated MIL and has a stored DTC P0420 – catalyst efficiency. Technician A says to check the engine for oil consumption. Technician B says to replace the downstream O_2 sensor. Who is correct?

TASK E.7

   A. A only

   B. B only

   C. Both A and B

   D. Neither A nor B

Answer A is correct. Only Technician A is correct. After a catalytic converter has been diagnosed faulty, faults that can cause catalyst damage, such as oil consumption, should be diagnosed.

Answer B is incorrect. No component should be replaced until it has been diagnosed faulty.

Answer C is incorrect. Only Technician A is correct.

Answer D is incorrect. Technician A is correct.

TASK F.5

30. A contaminated MAF sensor will increase which of the following emissions?

    A. HC

    B. CO_2

    C. CO

    D. O_2

    Answer A is incorrect. A contaminated MAF sensor will not cause a severe enough lean condition to cause increased HC emissions.

    Answer B is incorrect. CO_2 levels will decrease at any air/fuel ratio other than 14.7:1.

    Answer C is incorrect. CO levels increase under rich conditions.

    Answer D is correct. A contaminated MAF sensor will create a lean condition. O_2 levels will increase when lean conditions exist.

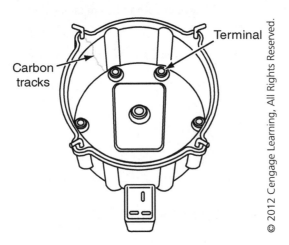

TASK C.12

31. The carbon track in the distributor cap pictured above could be caused by:

    A. A cracked spark plug

    B. High coil wire resistance

    C. High plug wire resistance

    D. Insufficient plug gap

    Answer A is incorrect. A cracked spark plug would create a low resistance path to ground and would not create a carbon track inside the distributor cap.

    Answer B is incorrect. High coil wire resistance would cause damage to the ignition coil. Secondary voltage would seek an easier path to ground before it ever got to the distributor cap.

    Answer C is correct. High resistance in a spark plug wire can cause voltage to seek another path to ground inside the distributor cap. This can cause a carbon track inside the cap.

    Answer D is incorrect. Insufficient spark plug gap would create a lower resistive path to ground and would not cause a carbon track to form.

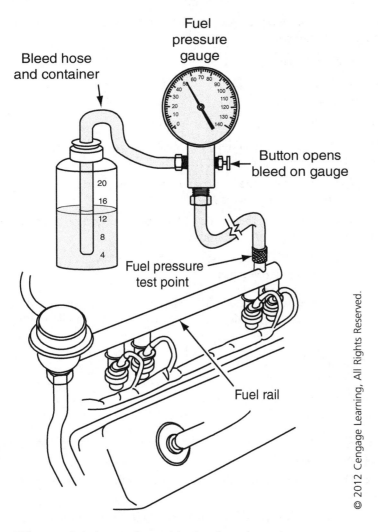

32. What test is being performed in the above image?

    A. Fuel pump pressure

    B. Fuel pump volume

    C. Fuel injector balance

    D. Fuel injector flow

TASK D.12

Answer A is incorrect. The fuel is being let out of the line, so pressure is not being measured. The pressure shown on the gauge would not be accurate if fuel were allowed to escape from the gauge.

Answer B is correct. The fuel is being directed into the container so that the volume of fuel flow can be measured.

Answer C is incorrect. Performing an injector balance test does not require directing fuel into a container. It would involve using an injector balance tester while watching the pressure on the gauge.

Answer D is incorrect. Injector flow is not being tested. Injectors must be removed and installed onto special equipment to test their flow rate.

TASK B.19

33. The TPS voltage is displayed as 0.5 V on the scan tool but measures 1.0 V at the sensor using a DVOM. The cause of this could be:

    A. High TPS signal wire resistance
    B. Low TPS reference voltage
    C. High TPS signal return resistance
    D. Low TPS sensor resistance

    Answer A is correct. Excessive resistance in the TPS signal wire can cause the voltage received at the ECM to be lower than what is sent from the TPS.

    Answer B is incorrect. The TPS signal at the sensor and what is displayed on the scan tool would be the same if the reference voltage were low.

    Answer C is incorrect. The TPS signal at the sensor and what is displayed on the scan tool would be the same if the signal return wire had high resistance.

    Answer D is incorrect. The TPS signal at the sensor and what is displayed on the scan tool would be the same if the TPS had low resistance.

TASK E.9

34. During a test of an electric air injection reaction (AIR) system, the pump is actuated with the scan tool and air is injected upstream. What does a low O_2 sensor voltage indicate under these conditions?

    A. A properly-operating system
    B. A malfunctioning diverter valve
    C. A malfunctioning AIR pump
    D. A malfunctioning O_2 sensor

    Answer A is correct. When extra oxygen is being injected into the exhaust, the upstream O_2 sensor should show a low voltage.

    Answer B is incorrect. The O_2 sensor would show no reaction if the diverter valve were malfunctioning.

    Answer C is incorrect. The O_2 sensor would show no reaction if the AIR pump were malfunctioning.

    Answer D is incorrect. The O_2 sensor is displaying the proper reading for the conditions that are present.

TASK F.9

35. Excessive no-load CO emissions would LEAST LIKELY be caused by:

    A. A clogged PCV valve
    B. A leaking fuel injector
    C. Excessive fuel pressure
    D. A fouled spark plug

    Answer A is incorrect. Since much of the idle air required by the engine is received through the PCV system, a clogged PCV valve will cause a lack of air and a rich condition, which will raise CO emissions.

    Answer B is incorrect. A leaking fuel injector will cause a rich condition and elevated CO emissions.

    Answer C is incorrect. Excessive fuel pressure would cause more fuel to be injected into the engine, which would increase CO emissions, especially at idle.

    Answer D is correct. A fouled spark plug will cause increased HC emissions. Since no fuel is burned, CO cannot form.

36. All of the following should be done while preparing to reprogram an ECM EXCEPT:

    A. Turn on the radio
    B. Test the battery and generator
    C. Disable daytime running lamps
    D. Ensure scan tool communication

TASK B.6

Answer A is correct. The radio should not be turned on, as this could drain the battery and interrupt the programming procedure.

Answer B is incorrect. A complete test of the battery and charging system should be done before reprogramming any ECM.

Answer C is incorrect. Leaving the daytime running lamps on could cause battery voltage to lower and disrupt the programming process.

Answer D is incorrect. A lack of scan tool communication can cause problems while reprogramming.

37. Excessive loaded HC emissions can be caused by which of the following?

    A. A clogged fuel filter
    B. A clogged air filter
    C. High plug wire resistance
    D. High fuel pressure

TASK F.10

Answer A is incorrect. A clogged fuel filter will cause a lean condition, which will result in increased O_2 levels.

Answer B is incorrect. A clogged air filter may cause elevated CO levels.

Answer C is correct. High plug wire resistance will cause the engine to miss under a loaded condition, increasing HC emissions.

Answer D is incorrect. High fuel pressure may cause increased CO emissions, especially at idle.

38. All of the following are symptoms of a sluggish O_2 sensor EXCEPT:

    A. Reduced fuel economy
    B. Poor idle quality
    C. Hesitation on acceleration
    D. Lower exhaust emissions

TASK B.11

Answer A is incorrect. A sluggish O_2 sensor can cause reduced fuel economy due to an incorrect air/fuel ratio.

Answer B is incorrect. A sluggish O_2 sensor can cause the idle quality of the vehicle to suffer, since the air/fuel ratio will be incorrect.

Answer C is incorrect. A sluggish O_2 sensor can cause the vehicle to hesitate.

Answer D is correct. A sluggish O_2 sensor will increase exhaust emissions.

TASK A.7

39. The composite vehicle has entered the shop with the MIL lamp illuminated and DTCs P0014 and P0021 – intake valve timing control fault banks 1 and 2. Technician A says that this could be caused by an open cam position solenoid. Technician B says that this could be caused by low oil pump output. Who is correct?

   A. A only

   B. B only

   C. Both A and B

   D. Neither A nor B

   Answer A is incorrect. An open cam position solenoid would only set a code for one bank. Both banks are affected in this situation.

   Answer B is correct. Only Technician B is correct. Since the camshaft actuators rely on oil pressure for camshaft positioning, if oil pump output were low there may not be enough pressure to move the actuators.

   Answer C is incorrect. Only Technician B is correct.

   Answer D is incorrect. Technician B is correct.

TASK E.12

40. A stuck open PCV valve could cause all of the following EXCEPT:

   A. Increased engine idle speed

   B. Decreased fuel trim readings

   C. Rough idle concerns

   D. Multiple cylinder misfire

   Answer A is incorrect. The excess air in the engine will cause higher-than-normal engine idle speeds.

   Answer B is correct. A stuck open PCV valve is more likely to cause increased fuel trim readings. False air may be drawn into the crankcase and cause a lean condition.

   Answer C is incorrect. A MAF controlled vehicle could have a rough idle if the PCV system is allowing false air into the system.

   Answer D is incorrect. The lean condition caused by excessive air could cause multiple cylinders to misfire.

TASK B.4

41. The throttle system is being diagnosed on the composite vehicle. Technician A says that two TP sensors that display identical voltages are used in the event that one fails. Technician B says that two accelerator pedal position (APP) sensors that display identical voltages are used in the event that one fails. Who is correct?

   A. A only

   B. B only

   C. Both A and B

   D. Neither A nor B

   Answer A is incorrect. The TAC system uses two TP sensors; however, they do not display identical voltage signals. Voltage increases for sensor one as the throttle opens, and voltage decreases for sensor two as the throttle opens.

   Answer B is incorrect. The TAC system uses two APP sensors, but they do not display identical voltage signals. Sensor one varies from 0.5 V at closed throttle to 3.5 V at wide-open throttle (WOT.) Sensor two varies from 1.5 V at closed throttle to 4.5 V at WOT.

   Answer C is incorrect. Neither Technician is correct.

   Answer D is correct. Neither Technician is correct. All TAC systems use redundant sensors in the event that one fails. These sensors rarely read the same voltages and certainly do not on the composite vehicle, as indicated by the composite vehicle manual.

42. A vacuum leak at the injector o-ring could cause any of these EXCEPT:

    A. Black exhaust smoke
    B. Low O_2 sensor voltage
    C. Rough unstable idle
    D. Spark knock

    TASK A.13

    Answer A is correct. Black exhaust smoke is caused by a rich condition. A leaking injector o-ring would cause a lean condition.

    Answer B is incorrect. A leaking injector o-ring would cause a lean condition, which would cause O_2 sensor voltage to be low.

    Answer C is incorrect. A leaking injector o-ring would cause a rough, unstable idle, since the cylinder would be lean and possibly misfire.

    Answer D is incorrect. The lean condition created by the vacuum leak may cause that cylinder to spark knock.

43. The composite vehicle has DTC P0301 – cylinder #1 misfire. The ignition coil for cylinder one has no spark. An oscilloscope is used to measure the voltage at terminal 14 of the ECM and displays steady system voltage. What component has failed?

    A. The ignition coil
    B. The spark plug
    C. The ECM
    D. The plug wire

    TASK C.10

    Answer A is incorrect. If the coil had failed, the measured voltage would be 0 V for an open primary circuit or the voltage would be falling to 0 V each time that the ECM commanded the coil to fire.

    Answer B is incorrect. The measured voltage would fluctuate between 0 V and system voltage at a regular interval if the spark plug were faulty.

    Answer C is correct. The voltage is remaining the same at all times, indicating that the ECM is not commanding the coil to fire.

    Answer D is incorrect. The composite vehicle does not use spark plug wires.

44. The composite vehicle is brought into the shop with a low power concern and a fuel pressure measurement of 38 psi (262 kPa). Technician A says that directing the return line into a container while actuating the fuel pump will test the pump output. Technician B says that the fuel filter should be checked for a restriction. Who is correct?

    A. A only
    B. B only
    C. Both A and B
    D. Neither A nor B

    TASK D.6

    Answer A is incorrect. The composite vehicle does not have a return-type fuel system.

    Answer B is correct. Only Technician B is correct. When fuel pressure is low, the fuel filter should be checked for a restriction.

    Answer C is incorrect. Only Technician B is correct.

    Answer D is incorrect. Technician B is correct.

TASK B.10

45. The composite vehicle's MAP sensor is reading 2.5 V at idle. This could be caused by:

    A. A restricted air filter

    B. A carbon-clogged MAP sensor

    C. An open MAP sensor signal wire

    D. A restricted exhaust

Answer A is incorrect. A restricted air filter would cause MAP sensor voltage to be low at idle and even lower when raising the RPM.

Answer B is incorrect. If the MAP sensor were clogged with carbon, it would read 4.5 V at all times.

Answer C is incorrect. An open MAP sensor wire would cause a signal voltage of 0 V.

Answer D is correct. A MAP voltage of 2.5 V correlates to 12 in. Hg. (60.7 kPa). This could be caused by a restricted exhaust.

TASK E.3

46. The composite vehicle enters the shop with a DTC P0452 – fuel tank pressure sensor low voltage. Which of the following could cause this condition?

    A. An open at ECM terminal 104

    B. An open at ECM terminal 110

    C. A short to ground at ECM terminal 104

    D. A short to ground at ECM terminal 110

Answer A is incorrect. An open at ECM terminal 104 would cause the EVAP vent solenoid to remain open at all times and set an EVAP leak code.

Answer B is incorrect. An open at ECM terminal 110 would cause the EVAP purge valve to be inoperative and cause an EVAP no purge flow code.

Answer C is correct. If ECM terminal 104 were shorted to ground, the EVAP vent solenoid would be closed at all times. This would cause a vacuum to form in the tank and lower fuel tank pressure sensor voltage.

Answer D is incorrect. A short to ground at ECM terminal 110 would cause the EVAP purge valve to be open at all times. This would cause the engine to idle roughly and could set an EVAP flow during non-purge code.

TASK D.11

47. The composite vehicle will not start and the fuel pressure (FP) sensor reads 0.5 V. There are no DTCs present. This could be caused by:

    A. A blown fuse #22

    B. An open at terminal 660 of the FPCM

    C. A faulty fuel pump

    D. A faulty fuel pressure (FP) sensor

Answer A is incorrect. If fuse #22 were blown, the fuel pump control module (FPCM) would not function and communication DTCs would set in the ECM.

Answer B is incorrect. Terminal 660 is the ground circuit for the fuel pump control module (FPCM). An open at this terminal would cause the module not to function and set communication DTCs in the ECM.

Answer C is correct. A reading of 0.5 V on the fuel pressure sensor indicates that there is no fuel pressure. This could be caused by a faulty fuel pump.

Answer D is incorrect. If the fuel pressure (FP) sensor were faulty, it would not prevent the engine from starting.

48.   The composite vehicle will not start. Technician A says that the resistance of the crank sensor cannot be measured and that waveform diagnosis should be used. Technician B says that the crank sensor will produce a square wave signal. Who is correct?

TASK C.5

A.   A only

B.   B only

C.   Both A and B

D.   Neither A nor B

Answer A is incorrect. Even though a resistance spec for the crank sensor is not given, the crank sensor is a magnetic-type sensor, and its resistance can be measured.

Answer B is incorrect. Since the sensor is a magnetic-type sensor, it will create a saw-tooth analog signal.

Answer C is incorrect. Neither Technician is correct.

Answer D is correct. Neither Technician is correct. The crank sensor on the composite vehicle can be checked for resistance and should create a sine wave pattern when functioning correctly.

49.   The composite vehicle has failed an I/M idle test for excessive CO levels. Technician A says that this could be caused by a leaking fuel pressure regulator. Technician B says that this could be caused by putting too much fuel in the fuel tank. Who is correct?

TASK F.4

A.   A only

B.   B only

C.   Both A and B

D.   Neither A nor B

Answer A is incorrect. The composite vehicle does not have a return-type fuel pressure regulator. The regulator is in the fuel tank. If it were leaking, it would cause low fuel pressure.

Answer B is incorrect. Typically, overfilling the fuel tank can saturate the charcoal canister and lead to a rich condition, but the composite vehicle has a vapor control valve that prevents liquid fuel from entering the charcoal canister.

Answer C is incorrect. Neither Technician is correct.

Answer D is correct. Neither Technician is correct. The composite vehicle likely failed because of a rich condition, such as a leaking fuel injector.

50.   The EGR valve position sensor on the composite vehicle is being tested KOEO. With the valve in the fully open position, the sensor should display what voltage reading?

TASK E.9

A.   0.5 V

B.   1.0 V

C.   4.0 V

D.   4.5 V

Answer A is incorrect. A voltage of 0.5 V indicates that the EGR valve is fully closed.

Answer B is incorrect. A voltage of 1.0 V indicates that the EGR valve is open approximately 11 percent.

Answer C is incorrect. A voltage of 4.0 V indicates that the EGR valve is open approximately 90 percent.

Answer D is correct. The EGR valve position sensor will display a reading of 4.5 V when fully open.

PREPARATION EXAM 5—ANSWER KEY

1.	D	21.	A	41.	D
2.	A	22.	B	42.	A
3.	C	23.	D	43.	C
4.	D	24.	C	44.	A
5.	B	25.	D	45.	B
6.	D	26.	A	46.	D
7.	A	27.	B	47.	A
8.	C	28.	A	48.	D
9.	C	29.	B	49.	B
10.	B	30.	C	50.	C
11.	A	31.	C		
12.	B	32.	D		
13.	D	33.	B		
14.	B	34.	A		
15.	A	35.	D		
16.	C	36.	B		
17.	D	37.	A		
18.	B	38.	C		
19.	A	39.	C		
20.	C	40.	A		

PREPARATION EXAM 5—EXPLANATIONS

TASK B.11

1. A throttle position sensor (TPS) is fixed at 4.8 V. Technician A says that the electronic control module (ECM) will ignore this reading and illuminate the MIL. Technician B says that the ECM will rely on other engine sensors to perform engine control operations. Who is correct?

A. A only

B. B only

C. Both A and B

D. Neither A nor B

Answer A is incorrect. A TPS reading of 4.8 V is a valid and recognizable reading from the TPS and will not be ignored or considered a fault.

Answer B is incorrect. When the ECM sees a TPS voltage this high, the vehicle will enter clear flood mode and not start.

Answer C is incorrect. Neither Technician is correct.

Answer D is correct. Neither Technician is correct. A TPS voltage of over 4.5 V will typically cause the ECM to enter clear flood mode, which will turn the injectors off. The reading will not be ignored, since it could be considered a valid TPS reading.

2. The engine coolant temperature (ECT) sensor reads erratic after the engine has warmed up. Technician A says that there may be air in the cooling system. Technician B says that ignition cables may be routed near the signal wire, giving a false reading. Who is correct?

TASK A.8

   A. A only

   B. B only

   C. Both A and B

   D. Neither A nor B

Answer A is correct. Only Technician A is correct. Air in the cooling system creates hotter pockets of air that will cause the sensor to read erratically.

Answer B is incorrect. While the ignition cables may induce a voltage into the ECT signal wire, this condition would occur all of the time instead of after warm-up.

Answer C is incorrect. Only Technician A is correct.

Answer D is incorrect. Technician A is correct.

SCAN TOOL DATA

Engine Coolant Temperature Sensor (ECT) 0.46 Volts	Intake Air Temperature Sensor (IAT) 2.24 Volts	MAP Sensor 1.8 V MAF Sensor 1.1 V	Throttle Position Sensor (TPS) 0.6 Volts
Engine Speed Sensor (RPM) 1500 rpm	Heated O₂ HO₂S Upstream 0.2–0.5 V Downstream 0.1–0.3 V	Vehicle Speed Sensor (VSS) 0 mph	Battery Voltage (B+) 14.2 Volts
Idle Air Control Valve (IAC) 0 percent	Evaporative Emission Canister Solenoid (EVAP) OFF	Torque Converter Clutch Solenoid (TCC) OFF	EGR Valve Control Solenoid (EGR) 0 percent
Malfunction Indicator Lamp (MIL) OFF	Diagnostic Trouble Codes NONE	Open/Closed Loop CLOSED	Fuel Pump Relay (FP) On
Fuel Level Sensor 3.5 V	Fuel Tank Pressure Sensor 2.0 V	Transmission Fluid Temperature Sensor 0.6 V	Transmission Turbine Shaft Speed Sensor 0 mph
Transmission Range Switch PARK	Trans Pressure Control Solenoid 80%	Transmission Shift Solenoid 1 ON	Transmission Shift Solenoid 2 OFF

Measured Ignition Timing °BTDC   Base Timing: 10°   Actual Timing: 20°

3. What diagnosis can be made using the scan tool data in the above figure?

TASK B.9

   A. A leaking MAP sensor vacuum hose

   B. A faulty IAC valve

   C. An intake manifold vacuum leak

   D. An exhaust manifold leak

Answer A is incorrect. A leaking MAP sensor vacuum hose would cause a rich condition, and the O₂ sensor voltage would be high.

Answer B is incorrect. The IAC is being commanded closed because of the high idle speed.

Answer C is correct. The low IAC count, low O₂ sensor voltage, high idle, and high MAP sensor reading all point toward a vacuum leak.

Answer D is incorrect. An exhaust leak would cause the lean O₂ readings but not the other irregular data.

TASK C.11

4. How will the ECM react to high knock sensor voltage?

   A. Increase the coil's dwell

   B. Advance ignition timing

   C. Decrease the coil's dwell

   D. Retard ignition timing

Answer A is incorrect. The ECM determines the coil's dwell based on engine load conditions, not knock sensor voltage.

Answer B is incorrect. High knock sensor voltage is an indicator that ignition timing is too far advanced. Advancing ignition timing further will cause the engine to ping or spark knock even more.

Answer C is incorrect. High knock sensor voltage indicates that ignition timing needs to be retarded. This will reduce spark knock and pinging.

Answer D is correct. High knock sensor voltage indicates that ignition timing needs to be retarded.

TASK F.7

5. Which of the following tests would be the most beneficial if a vehicle had failed the I/M test for excessive CO emissions?

   A. Power balance test

   B. Fuel pressure test

   C. Exhaust backpressure

   D. Compression test

Answer A is incorrect. A power balance test would be appropriate if HC levels are high due to a cylinder misfire.

Answer B is correct. Excessive CO emissions indicate a rich condition, which could be caused by excessive fuel pressure.

Answer C is incorrect. Excessive CO emissions are an indicator of a rich condition, which cannot be caused by high backpressure.

Answer D is incorrect. Low compression can cause high HC emissions through problems such as a burned valve.

TASK E.8

6. Which of the following symptoms could LEAST LIKELY be associated with an EGR valve stuck open?

   A. Hard starting

   B. Rough idle

   C. Engine stalling

   D. Detonation

Answer A is wrong. A stuck open EGR valve causes hard engine starting due to a lack of oxygen.

Answer B is wrong. A stuck open EGR valve will cause rough idle, because there will be too much exhaust gas recirculating into the engine.

Answer C is wrong. A stuck open EGR valve will cause engine stalling, since there will not be enough oxygen for combustion..

Answer D is correct. The EGR prevents detonation by lowering cylinder temperatures. Detonation occurs when heat ignites the air/fuel mixture. If the EGR valve sticks open, detonation is not likely to occur.

7. When the ECM receives a low voltage signal from the HO$_2$S sensor it will:

    A. Increase fuel delivery
    B. Decrease fuel delivery
    C. Increase engine RPM
    D. Decrease engine RPM

TASK D.5

Answer A is correct. A low voltage from the HO$_2$S indicates a lean air/fuel ratio. To maintain a 14.7:1 air/fuel ratio, the ECM will have to increase fuel delivery by increasing the fuel injector pulse width (PW).

Answer B is incorrect. A low voltage from the HO$_2$S indicates a lean air/fuel ratio. Decreasing fuel delivery will cause an even leaner air/fuel ratio.

Answer C is incorrect. HO$_2$S voltage has little influence upon engine RPM.

Answer D is incorrect. HO$_2$S voltage has little influence upon engine RPM.

8. A faulty catalytic converter has been diagnosed. Technician A says that a malfunctioning PCV system could have caused the converter to fail. Technician B says that leaking valve stem seals could have caused the converter to fail. Who is correct?

    A. A only
    B. B only
    C. Both A and B
    D. Neither A nor B

TASK A.12

Answer A is incorrect. Technician B is also correct.

Answer B is incorrect. Technician A is also correct.

Answer C is correct. Both Technicians are correct. Both a malfunctioning PCV system and leaking valve guide seals can cause excessive oil consumption that can damage the catalytic converter.

Answer D is incorrect. Both Technicians are correct.

TASK D.11

9. A customer is concerned with a "rotten egg" smell coming from the exhaust, with no other drivability concerns. What should the technician do to correct this problem?

    A. Replace the catalytic converter

    B. Replace the O_2 sensors

    C. Suggest an alternate fuel source

    D. Suggest a complete tune-up

    Answer A is incorrect. Replacement of the catalytic converter will not correct this condition. Other conditions are causing the converter to produce the "rotten egg" smell.

    Answer B is incorrect. Replacement of the O_2 sensors will not correct this condition. The O_2 sensors can in no way cause the "rotten egg" smell.

    Answer C is correct. Sulfur content in fuel can vary among fueling stations. If fuel with high sulfur content is used, a "rotten egg" odor may come from the exhaust and may not indicate that there are other engine concerns. Replacement of components may not correct this condition.

    Answer D is incorrect. While a tune-up is good preventative maintenance, it will not correct this condition. A technician should never sell a tune-up without knowing if it will correct the customer's concern.

TASK F.5

10. A vehicle is brought into the shop with a DTC P0455 – EVAP system large leak detected. Technician A says to first check the gas cap to see if it is tight. Technician B says that a lack of vacuum to the purge valve can cause a leak detection. Who is correct?

    A. A only

    B. B only

    C. Both A and B

    D. Neither A nor B

    Answer A is incorrect. The system should be leak tested to verify that the fuel cap is actually leaking. Simply turning the cap and clearing the code could result in a comeback.

    Answer B is correct. Only Technician B is correct. A lack of vacuum to the purge valve will cause no vacuum to build in the fuel tank during the EVAP monitor. The ECM may interpret this as a large leak.

    Answer C is incorrect. Only Technician B is correct.

    Answer D is incorrect. Technician B is correct.

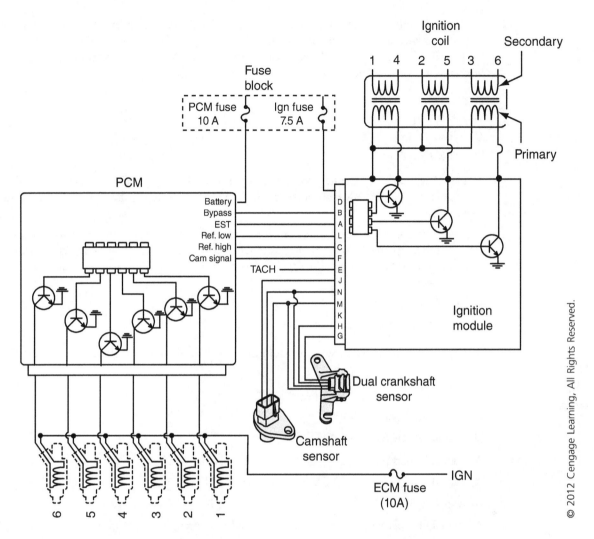

TASK C.8

11. The vehicle using the engine control system above will not start. The vehicle has spark but no injector pulse. Technician A says that the camshaft position (CMP) sensor could be faulty. Technician B says that the crankshaft position (CKP) sensor could be faulty. Who is correct?

A. A only

B. B only

C. Both A and B

D. Neither A nor B

Answer A is correct. Only Technician A is correct. The ECM needs the CMP to determine when to pulse the injectors, since this is an SFI system. The ignition module only needs the CKP to fire the coils, since it is a waste spark system.

Answer B is incorrect. If the crank sensor were faulty, there would be no spark.

Answer C is incorrect. Only Technician A is correct.

Answer D is incorrect. Technician A is correct.

TASK B.5

12. The oxygen sensor heater is being tested. Technician A says that measuring the voltage drop of the heater will verify its operation. Technician B says that when it is cold, a functioning heater will cause O₂ sensor voltage to drop below 100 mV when it reaches operating temperature without the engine running. Who is correct?

    A.   A only

    B.   B only

    C.   Both A and B

    D.   Neither A nor B

Answer A is incorrect. A voltage drop test will not verify the operation of the heater. Since the heater will be the only load in the circuit, it will drop system voltage even if it has high or low resistance.

Answer B is correct. Only Technician B is correct. Without the engine running, most of the air in the exhaust stream is ambient air containing high amounts of oxygen. As the sensor heats up, the voltage will continue to drop as it measures the high oxygen content in the exhaust system.

Answer C is incorrect. Only Technician B is correct.

Answer D is incorrect. Technician B is correct.

TASK E.10

13. A failure of which emission system would cause increased NO_x emissions?

    A.   Evaporative emissions (EVAP)

    B.   Positive crankcase ventilation (PCV)

    C.   Early fuel evaporation (EFE)

    D.   Exhaust gas recirculation (EGR)

Answer A is incorrect. The EVAP system controls HC emissions.

Answer B is incorrect. The PCV system controls HC emissions.

Answer C is incorrect. The EFE system controls HC and CO emissions.

Answer D is correct. The EGR system is used to control NO_x emissions.

TASK B.8

14. A rich O₂ sensor voltage would be:

    A.   Below 450 mV

    B.   Above 450 mV

    C.   Between 250 mV and 750 mV

    D.   Fixed at 500 mV

Answer A is incorrect. An O₂ sensor voltage less than 450 mV is considered to be a lean condition.

Answer B is correct. An O₂ sensor voltage greater than 450 mV is considered to be a rich reading.

Answer C is incorrect. An O₂ sensor voltage between 250 mV and 750 mV is considered to be a transition stage between rich and lean.

Answer D is incorrect. Any time that the O₂ sensor voltage is fixed at any voltage, a problem exists.

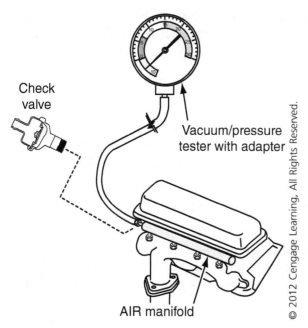

Check valve

Vacuum/pressure tester with adapter

AIR manifold

15. What is being tested in the above picture?

   A. Exhaust backpressure

   B. Check valve function

   C. AIR pump pressure

   D. EGR system flow

TASK A.5

Answer A is correct. Connecting a pressure gauge to the air injection distribution manifold will measure backpressure.

Answer B is incorrect. The check valve is tested by blowing air through it. Air should flow in only one direction.

Answer C is incorrect. Since the pressure gauge is on the exhaust manifold side, it would not measure AIR pump pressure. Exhaust backpressure would be measured instead.

Answer D is incorrect. EGR system flow cannot be measured in this manner, since the EGR gas would not flow through this portion of the exhaust.

16. A distributor ignition (DI) vehicle is brought into the shop with a DTC P0340 – CMP sensor/CKP sensor correlation. Technician A says that this can be caused by a worn distributor gear. Technician B says that this could be caused by worn cam timing components. Who is correct?

   A. A only

   B. B only

   C. Both A and B

   D. Neither A nor B

TASK C.9

Answer A is incorrect. Technician B is also correct.

Answer B is incorrect. Technician A is also correct.

Answer C is correct. Both Technicians are correct. A worn distributor gear will cause the cam sensor signal to be retarded and not be in synch with the crank sensor. Worn timing components, such as a tensioner or chain, can cause the cam sensor to be retarded as well.

Answer D is incorrect. Both Technicians are correct.

TASK B.19

17. The O_2 sensor voltage on a vehicle is fixed at 0.03 V. When the exhaust gas is measured directly with an exhaust gas analyzer, what gas will be high if the O_2 sensor voltage is correct?

A. CO

B. NO_x

C. CO_2

D. O_2

Answer A is incorrect. The O_2 sensor is reading lean. CO is a rich indicator.

Answer B is incorrect. NO_x levels do not directly coordinate with air/fuel ratio.

Answer C is incorrect. CO_2 levels will lower when running lean.

Answer D is correct. A low voltage from the O_2 sensor indicates a lean condition. O_2 levels will be high if the vehicle is running lean.

TASK D.1

18. A vehicle is hard to start on cold mornings after sitting all night. Technician A says that the vehicle will have to sit overnight before diagnosis can begin. Technician B says that the fuel may pressure should be monitored for any pressure loss when the vehicle is pulled into the service bay. Who is correct?

A. A only

B. B only

C. Both A and B

D. Neither A nor B

Answer A is incorrect. A leaking fuel pump check valve would cause hard starting regardless of temperature.

Answer B is correct. Only Technician B is correct. High alcohol content in the fuel can lead to hard cold starting.

Answer C is incorrect. Only Technician B is correct.

Answer D is incorrect. Technician B is correct.

TASK F.12

19. A vehicle fails an emission test for excessive NO_x at 2500 rpm. Technician A says that this could be caused by clogged EGR passages. Technician B says that this could be caused by a diverter valve stuck in the downstream position. Who is correct?

A. A only

B. B only

C. Both A and B

D. Neither A nor B

Answer A is correct. Only Technician A is correct. Clogged EGR passages would prevent exhaust gas from entering the combustion chamber, which would increase NO_x emissions.

Answer B is incorrect. The purpose of the air injection system is to add oxygen to aid in the reduction of HC and CO. NO_x emissions would not increase if the diverter valve were stuck in the downstream position.

Answer C is incorrect. Only Technician A is correct.

Answer D is incorrect. Technician A is correct.

20. The hoses of an air injection reaction (AIR) system show signs of burning. This could be a result of:

    A. A faulty air pump
    B. A faulty diverter valve
    C. A faulty check valve
    D. A restricted exhaust

    TASK E.14

    Answer A is incorrect. A faulty air pump will not cause the hoses to burn. The system would simply be inoperative if the air pump were faulty.

    Answer B is incorrect. A faulty diverter valve will cause the air to be injected either upstream or downstream at the wrong time, but it will not cause the hoses to burn.

    Answer C is correct. A faulty check valve will allow exhaust gas to enter the AIR distribution hoses, which are not designed to carry exhaust gas. The heat will cause them to burn.

    Answer D is incorrect. The check valve should not allow exhaust gas to get past the check valve, even if the exhaust is restricted. It should be strong enough to prevent exhaust gas flow into the AIR system, even if excessive backpressure exists.

21. Dielectric grease is used on many connectors and terminals when a vehicle's electrical system is being serviced. Technician A says that the grease will help keep water out of the connectors. Technician B says that the grease will allow for a better connection because it is conductive. Who is correct?

    A. A only
    B. B only
    C. Both A and B
    D. Neither A nor B

    TASK B.10

    Answer A is correct. Only Technician A is correct. Dielectric grease is used to seal connectors and keep water out of the connector.

    Answer B is incorrect. Dielectric grease is not conductive and will not increase continuity.

    Answer C is incorrect. Only Technician A is correct.

    Answer D is incorrect. Technician A is correct.

22. A burned exhaust valve is likely to increase which two of the following emissions?

    A. CO_2 and CO
    B. HC and O_2
    C. NO_x and CO
    D. HC and CO

    TASK A.9

    Answer A is incorrect. CO_2 and CO levels will lower due to the lack of combustion.

    Answer B is correct. HC and O_2 will leak out of the cylinder and into the exhaust stream.

    Answer C is incorrect. A burned valve will not affect NO_x levels. CO levels will go down.

    Answer D is incorrect. HCs will increase, but CO will not.

SCAN TOOL DATA

Engine Coolant Temperature Sensor (ECT)	Intake Air Temperature Sensor (IAT)	MAP Sensor 4.5 V MAF Sensor 0.2 V	Throttle Position Sensor (TPS)
0 Volts	2.9 Volts		0.6 Volts
Engine Speed Sensor (RPM)	Heated O_2 HO_2S Upstream 0 V Downstream 0 V	Vehicle Speed Sensor (VSS)	Battery Voltage (B+)
0 rpm		0 mph	12.6 Volts
Idle Air Control Valve (IAC)	Evaporative Emission Canister Solenoid (EVAP)	Torque Converter Clutch Solenoid (TCC)	EGR Valve Control Solenoid (EGR)
30 percent	OFF	OFF	0 percent
Malfunction Indicator Lamp (MIL)	Diagnostic Trouble Codes	Open/Closed Loop	Fuel Pump Relay (FP)
ON	P0117	OPEN	OFF
Fuel Level Sensor	Fuel Tank Pressure Sensor	Transmission Fluid Temperature Sensor	Transmission Turbine Shaft Speed Sensor
3.3 V	4.5 V	1.8 V	0 mph
Transmission Range Switch	Trans Pressure Control Solenoid	Transmission Shift Solenoid 1	Transmission Shift Solenoid 2
PARK	0	OFF	OFF

Measured Ignition Timing °BTDC   Base Timing: 10°   Actual Timing: 0

TASK B.8

23. Based on the KOEO scan tool data shown in the above figure, the DTC P0117 would LEAST LIKELY be caused by which of the following?

    A. ECT sensor shorted

    B. ECT sensor signal circuit shorted to ground

    C. A faulty ECM

    D. ECT sensor open

Answer A is incorrect. An internally shorted ECT sensor would cause ECT sensor voltage to drop to zero.

Answer B is incorrect. An open ECT sensor would display a voltage close to 5 V.

Answer C is incorrect. An ECM could short out internally and cause ECT voltage to drop to 0 V.

Answer D is correct. The reading of 0 V from the ECT sensor indicates that the sensor is shorted.

TASK F.5

24. Which of the following gases would be elevated if a rich condition were present?

    A. CO_2

    B. NO_x

    C. CO

    D. O_2

Answer A is incorrect. When the system is running efficiently, CO_2 levels should be high.

Answer B is incorrect. NO_x is not a rich indicator. NO_x is created from the nitrogen in the air combining with oxygen under high heat and pressure conditions.

Answer C is correct. Higher-than-normal CO levels indicate that the vehicle is running rich.

Answer D is incorrect. High O_2 levels indicate a lean condition.

25. A restricted catalytic converter is suspected on a vehicle, but the O_2 sensors are too difficult to reach in order to measure backpressure. Which of the following tools can be used to verify the restricted converter?

TASK A.5

   A. An oscilloscope

   B. A digital multi-meter (DMM)

   C. A compression gauge

   D. A vacuum gauge

   Answer A is incorrect. An oscilloscope cannot be used to diagnose a catalytic converter, because it will only show electrical signals, and the catalytic converter is a chemical device.

   Answer B is incorrect. A DMM cannot be used to diagnose a catalytic converter. DMMs measure electrical signals.

   Answer C is incorrect. A compression gauge cannot be used to diagnose a catalytic converter. A compression gauge is used to measure the engine's compression.

   Answer D is correct. While performing a snap test, engine vacuum should drop to 0 in. Hg. (0 cm. Hg.) when the throttle is opened and quickly climb to over 20 in. Hg. (50.7 cm. Hg) when the throttle is shut. If the exhaust is restricted, vacuum will slowly climb after the throttle is shut.

26. A vehicle cuts out and stumbles at times and has a DTC, P0122 – TPS low voltage stored in its history. Technician A says that the sensor voltage should be viewed with a scope. Technician B says that a sudden voltage drop to zero displayed on the scope indicates that the sensor is shorted. Who is correct?

TASK B.17

   A. A only

   B. B only

   C. Both A and B

   D. Neither A nor B

   Answer A is correct. Only Technician A is correct. A scope should be used to identify a bad spot in the TPS.

   Answer B is incorrect. A sudden drop in voltage displayed on the scope indicates that the sensor has an open, not a short.

   Answer C is incorrect. Only Technician A is correct.

   Answer D is incorrect. Technician A is correct.

27. A catalytic converter is being tested after replacement, and the inlet and outlet temperatures are measured. What should the temperatures show?

TASK F.17

   A. A hotter inlet temperature

   B. A hotter outlet temperature

   C. The same temperatures should be measured

   D. Measuring the temperature is irrelevant

   Answer A is incorrect. The outlet of the catalytic converter should be hotter than the inlet. This indicates that the converter is functioning.

   Answer B is correct. An operating catalytic converter will create heat while it is converting the gases. The outlet temperature should be higher than the inlet temperature.

   Answer C is incorrect. The catalytic converter is not functioning if the inlet and outlet temperatures are the same.

   Answer D is incorrect. Measuring the inlet and outlet temperatures of the catalytic converter is a simple and fairly effective method of determining its effectiveness.

TASK B.2

28. A common method of tampering involves installing a resistor in series with the intake air temperature (IAT) sensor. What affect will this have on the sensor voltage?

    A. IAT voltage will increase
    B. IAT voltage will decrease
    C. Sensor voltage will be bypassed
    D. Sensor voltage will remain steady and high

    Answer A is correct. The increased resistance in the IAT circuit will cause IAT voltage to increase.

    Answer B is incorrect. Sensor voltage would decrease if the resistor were installed in parallel with the IAT sensor.

    Answer C is incorrect. The sensor will not be bypassed by installing the resistor.

    Answer D is incorrect. Sensor voltage will be higher than normal but will still change with changes in temperature.

TASK D.5

29. The scan tool is displaying a long-term fuel trim of (LTFT) –25 percent. What is the ECM trying to compensate for?

    A. A lean condition
    B. A rich condition
    C. An advanced distributor
    D. A retarded distributor

    Answer A is incorrect. When a lean condition exists, the ECM will add fuel to the engine, and LTFT numbers will be shown positive.

    Answer B is correct. When LTFT is showing negative numbers, the ECM is taking fuel away from the engine to achieve a 14.7:1 air/fuel ratio.

    Answer C is incorrect. LTFT has nothing to do with ignition timing.

    Answer D is incorrect. LTFT has nothing to do with ignition timing.

TASK B.7

30. Which of the following would prevent the EVAP monitor from running?

    A. A leaking fuel cap
    B. Faulty purge valve
    C. Low fuel level
    D. A vacuum leak

    Answer A is incorrect. The EVAP monitor will run if the fuel cap is leaking; however, it will not pass.

    Answer B is incorrect. The EVAP monitor will run if the purge valve is faulty; however, it will not pass.

    Answer C is correct. The EVAP monitor will not run if the fuel level is lower than ¼ of a tank.

    Answer D is incorrect. The EVAP monitor will run if a vacuum leak is present. The location of the leak may cause the monitor to fail.

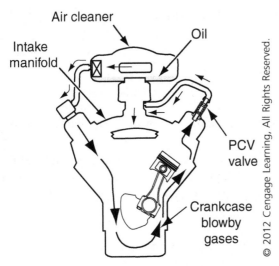

31. Oil is found in the air filter of the engine pictured above. Technician A says that the PCV valve may not be functional. Technician B says that the rings may be worn. Who is correct?

TASK E.8

A. A only

B. B only

C. Both A and B

D. Neither A nor B

Answer A is incorrect. Technician B is also correct.

Answer B is incorrect. Technician A is also correct.

Answer C is correct. Both Technicians are correct. A non-functioning PCV valve will cause crankcase pressure to build, and the easiest exit path is through the fresh air hose. Excessive wear of the piston rings would create more blowby gas than the PCV system can handle. In this case, the excess blowby would exit the fresh air hose.

Answer D is incorrect. Both Technicians are correct.

32. Refer to the following idle speed emissions levels to determine the possible fault:

  HC – 40 ppm

  CO – 4.3 percent

  CO_2 – 10 percent

  O_2 – 0.1 percent

TASK F.9

A. A restricted fuel filter

B. An intake vacuum leak

C. Insufficient fuel pressure

D. Stuck open EVAP purge valve

Answer A is incorrect. A restricted fuel filter would cause a lean condition. O_2 levels would increase if a lean condition were present.

Answer B is incorrect. An intake manifold vacuum leak would cause a lean condition. Lean conditions will not cause increased CO emissions.

Answer C is incorrect. Insufficient fuel pressure would cause a lean condition. While this will cause CO_2 levels to decrease, it would also cause the O_2 levels to increase.

Answer D is correct. The emissions readings shown indicate a rich condition. A stuck open EVAP purge valve will cause a rich condition, resulting in increased CO emissions.

TASK E.12

33. A stuck closed PCV valve could cause which of the following conditions?

　　A.　Excessive blowby

　　B.　Oil in the intake pipe

　　C.　High idle speed

　　D.　Lower exhaust emissions

Answer A is incorrect. The PCV system has nothing to do with the amount of blowby created by the engine.

Answer B is correct. Should the PCV valve get stuck closed, pressure caused by blowby gases may blow oil out of the PCV inlet, which is typically connected to the intake piping.

Answer C is incorrect. A stuck closed PCV valve will not cause a high idle speed.

Answer D is incorrect. A fault in any emission system will not lower exhaust emission.

TASK F.5

34. A stuck open thermostat will increase which of the following emissions?

　　A.　CO

　　B.　CO_2

　　C.　NO_x

　　D.　O_2

Answer A is correct. A stuck open thermostat will cause the engine to require more fuel, which will cause a rich condition, indicated by high CO emissions.

Answer B is incorrect. A stuck open thermostat will lower CO_2 levels.

Answer C is incorrect. A stuck open thermostat will not directly affect NO_x emissions, but overall the cooler engine conditions could lower NO_x emissions.

Answer D is incorrect. A stuck open thermostat will cause the engine to require a richer mixture, which will lower O_2 levels.

TASK B.21

35. All of the following can cause the O_2 sensor to become contaminated with oil EXCEPT:

　　A.　Leaking valve guides

　　B.　PCV system faults

　　C.　Worn piston rings

　　D.　Leaking rear main seal

Answer A is incorrect. Leaking valve guides, especially exhaust guides, would allow oil to enter the exhaust system. This would cause the O_2 sensor to become contaminated with oil.

Answer B is incorrect. PCV system faults could allow excessive oil to enter the combustion chamber and eventually contaminate the O_2 sensor with oil.

Answer C is incorrect. Worn piston rings would cause excessive oil to enter the combustion chamber and then enter the exhaust, which would contaminate the O_2 sensor.

Answer D is correct. A leaking rear main seal will not cause damage to the O_2 sensor.

36. A symptom of an air injection system directing air upstream at all times would be:

    A. Random cylinder misfire
    B. Increased fuel consumption
    C. Low power output
    D. Engine backfire on acceleration

TASK E.12

    Answer A is incorrect. Constant air injection would not cause a random misfire but would cause the O_2 sensors to constantly read lean.

    Answer B is correct. The injected air would cause the O_2 sensors to read lean. The ECM would then richen the air/fuel ratio, which would result in increased fuel consumption.

    Answer C is incorrect. Low power output would not be expected in this situation, as the engine would have plenty of fuel because of the lean O_2 signals.

    Answer D is incorrect. The engine may backfire on deceleration if air is constantly injected into the exhaust.

37. A vehicle failed the no-load I/M test for excessive HC emissions; however, it passed the loaded test for HCs. Which of the following could cause this condition?

    A. A vacuum leak
    B. A fouled spark plug
    C. Clogged EGR passages
    D. A clogged fuel filter

TASK F.8

    Answer A is correct. A vacuum leak would cause excessive emissions output at idle but would be less problematic at higher RPM.

    Answer B is incorrect. A fouled spark plug would cause high HC emissions at all times.

    Answer C is incorrect. Clogged EGR passages would affect NO_x emissions, not HC emissions.

    Answer D is incorrect. A clogged fuel filter would cause a lean condition, which would increase O_2 emissions.

38. A vehicle with distributorless electronic ignition (EI) has an engine miss under heavy load but runs okay at idle. All of the following could be the cause of the miss EXCEPT:

    A. A carbon tracked spark plug
    B. Excessive plug wire resistance
    C. A faulty ignition module
    D. A faulty ignition coil

TASK C.9

    Answer A is incorrect. A carbon tracked spark plug can cause the vehicle to miss under heavy load.

    Answer B is incorrect. Excessive plug wire resistance can cause the vehicle to miss under heavy load.

    Answer C is correct. The operation of the ignition module will not be affected by engine load.

    Answer D is incorrect. A faulty ignition coil can cause the vehicle to miss under heavy load.

TASK B.5

39. The composite vehicle enters the shop with a DTC P0031 – Bank 1 AFR Sensor heater circuit malfunction. Which of the following would be the best first step in diagnosis?

A. Replace the AFR Sensor
B. Check for a blown fuse 20
C. Check the heater resistance
D. Check for a damaged ECM terminal 152

Answer A is incorrect. A component should never be replaced without proper diagnosis.

Answer B is incorrect. If fuse 20 were blown, the vehicle would not start and multiple DTCs would set, not just a P0031.

Answer C is correct. Checking the sensor heater resistance is the best step for diagnosis.

Answer D is incorrect. ECM terminal 152 is for the sensor signal, not the heater circuit.

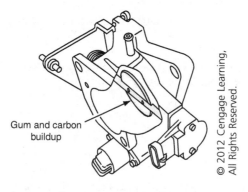

Gum and carbon buildup

© 2012 Cengage Learning, All Rights Reserved.

TASK D.10

40. The problem indicated in the figure above can cause all of the following problems EXCEPT:

A. High idle
B. Sticking throttle
C. Stalling
D. Rough idle

Answer A is correct. Gum and carbon on the throttle plate will restrict the amount of air entering the engine. This will not cause a high idle.

Answer B is incorrect. Gum and carbon buildup on the throttle body can cause a sticking throttle.

Answer C is incorrect. Gum and carbon buildup on the throttle body can cause stalling.

Answer D is incorrect. Gum and carbon buildup on the throttle body can cause a rough idle.

TASK E.8

41. The composite vehicle is slow to refuel. Technician A says that a clogged purge solenoid will cause this condition. Technician B says that a stuck open vent solenoid can cause this condition. Who is correct?

A. A only
B. B only
C. Both A and B
D. Neither A nor B

Answer A is incorrect. The purge solenoid is normally closed and will not cause the described symptoms.

Answer B is incorrect. The vent solenoid is normally open and will cause the described symptoms only if it is stuck closed.

Answer C is incorrect. Neither Technician is correct.

Answer D is correct. Neither Technician is correct. The purge valve should be closed, so if it is clogged, it will not cause the concern. The vent valve should be open, so if it is stuck open, it will not cause the concern either.

42. The composite vehicle has poor acceleration and an illuminated MIL. Technician A says that one failed TPS can cause this. Technician B says that an open at terminal "a" of the TPS will cause this condition. Who is correct?

TASK D.9

   A. A only

   B. B only

   C. Both A and B

   D. Neither A nor B

   Answer A is correct. Only Technician A is correct. Should one TPS fail, the TAC system will limit the throttle to 35 percent maximum opening. This will cause poor acceleration and an illuminated MIL.

   Answer B is incorrect. An open at terminal "a" of the TPS will cause the sensors to lose their 5 V reference and the vehicle to have no acceleration. The throttle will default to 15 percent open.

   Answer C is incorrect. Only Technician A is correct.

   Answer D is incorrect. Only Technician A is correct.

43. The composite vehicle has an illuminated MIL and DTCs P0011 and P0021 – intake valve timing control fault banks 1 and 2. This could be caused by all of the following EXCEPT:

TASK A.7

   A. Low oil level

   B. Incorrect oil viscosity

   C. A faulty cam sensor

   D. Low oil pump output

   Answer A is incorrect. Since the VVT system relies on oil pressure to rotate the cams, low oil level may affect the system operation.

   Answer B is incorrect. Since the VVT system relies on oil pressure to rotate the cams, the wrong oil may not properly rotate the cams.

   Answer C is correct. One single faulty cam sensor would not set this code for both banks.

   Answer D is incorrect. Since the VVT system relies on oil pressure to rotate the cams, low oil pump output could cause the system to malfunction.

44. Technician A says that if an invalid key is used to start the composite vehicle, the ECM will disable the fuel injectors within two seconds. Technician B says that if an invalid key is used to start the composite vehicle, the security indicator will remain on constantly. Who is correct?

TASK B.13

   A. A only

   B. B only

   C. Both A and B

   D. Neither A nor B

   Answer A is correct. Only Technician A is correct. Should an invalid key be used to start the composite vehicle, the ECM will disable the fuel injectors after two seconds, and the engine will die.

   Answer B is incorrect. Should an invalid key be used to start the composite vehicle, the security indicator will flash.

   Answer C is incorrect. Only Technician A is correct.

   Answer D is incorrect. Technician A is correct.

TASK C.8

45. Which of the following sensors, if faulty, will prevent the composite vehicle from starting?

  A. CMP sensor

  B. CKP sensor

  C. Knock sensor (KS)

  D. MAP sensor

Answer A is incorrect. If only one CMP sensor is faulty, the composite vehicle will still start. The ECM would simply look at the other CMP sensors and set a code for the malfunctioning sensor.

Answer B is correct. If CKP signal is not received by the ECM within two seconds after the ignition switch is turned on, the fuel pressure control module (FPCM) will be turned off and the vehicle will not start.

Answer C is incorrect. A faulty KS will not prevent the composite vehicle from starting. A DTC would be set for the knock sensor, and ignition timing advances would not occur.

Answer D is incorrect. A faulty MAP sensor will not prevent the composite vehicle from starting. A DTC would be set, and many monitors would not run, as they require the MAP sensor to run those tests.

TASK E.9

46. When testing the EVAP system on the composite vehicle, the fuel tank pressure sensor remains at 1 V and slowly increases after the vent and purge solenoids have been commanded off. This could indicate:

  A. A stuck closed purge solenoid

  B. A stuck open vent solenoid

  C. A clogged purge hose

  D. A clogged vent hose

Answer A is incorrect. If the purge solenoid were stuck closed, a vacuum would never reach the tank since the solenoid controls vacuum.

Answer B is incorrect. If the vent solenoid were stuck open, a vacuum would not be able to form inside the tank, because air would be able to enter the vent.

Answer C is incorrect. If the purge hose were clogged, a vacuum would not be able to reach the tank through the blocked hose.

Answer D is correct. A clogged vent hose will cause vacuum to remain inside the fuel tank when the vent is opened. This will cause the fuel tank pressure sensor voltage to remain low.

TASK B.24

47. After a repair, the keep alive memory (KAM) in the ECM can be reset by all of the following EXCEPT:

  A. Clearing ECM DTCs

  B. Removing the ECM fuse

  C. Unplugging the ECM

  D. Disconnecting the battery

Answer A is correct. Clearing DTCs will not always clear all data stored in the ECM's KAM. Some scan tools will allow the technician to clear the KAM, which is the preferred method.

Answer B is incorrect. Removing the ECM fuse will clear the ECM's KAM.

Answer C is incorrect. Unplugging the ECM will clear the ECM's KAM.

Answer D is incorrect. Disconnecting the battery will clear the ECM's KAM, but may cause other modules to lose critical data. This is the least preferred method of clearing the KAM.

48. The composite vehicle will not start, so the fuel pump enable circuit is being tested. While measuring the resistance from ECM terminal 127 to fuel pump control module (FPCM) terminal 649, the ohmmeter displays O.L. This reading indicates:

TASK D.6

    A. A good fuel pump enable circuit

    B. A bad fuel pump control module (FPCM)

    C. An open fuel pump

    D. A faulty fuel pump relay

Answer A is incorrect. A resistance reading of O.L indicates that the circuit is open.

Answer B is incorrect. This circuit is the signal from the ECM to the fuel pump control module (FPCM) so it can command the correct duty cycle for the fuel pump. Whether this circuit is good or not will not determine the state of the fuel pump control module.

Answer C is incorrect. This circuit is the signal from the ECM to the fuel pump control module (FPCM). It will not determine the condition of the fuel pump.

Answer D is correct. The composite vehicle does not utilize a fuel pump relay.

49. One ignition coil on the composite vehicle is suspected of causing a misfire. Which of the following would be an acceptable primary coil resistance reading?

TASK C.3

    A. $0.1\ \Omega$

    B. $1\ \Omega$

    C. $10\ K\Omega$

    D. O.L.

Answer A is incorrect. $0.1\ \Omega$ indicates that the primary coil is shorted.

Answer B is correct. $1\ \Omega$ is an acceptable reading for the primary ignition coil resistance.

Answer C is incorrect. $10\ K\Omega$ would be considered excessive for primary ignition coil resistance; however, this would be acceptable for the secondary coil

Answer D is incorrect. A reading of O.L. would indicate that the coil is open.

50. Which sensor on the composite vehicle is used primarily to detect exhaust flow in the EGR system?

TASK E.3

    A. EGR position

    B. Throttle position

    C. MAP sensor

    D. MAF sensor

Answer A is incorrect. While the EGR position sensor will tell if the EGR valve is open, it will not detect an exhaust gas flow problem.

Answer B is incorrect. The EGR system operation will have no effect on TPS readings.

Answer C is correct. The ECM uses the MAP sensor to detect changes in manifold pressure in order to verify EGR system flow.

Answer D is incorrect. The ECM does not use the MAF sensor to check the operation of the EGR system.

PREPARATION EXAM 6—ANSWER KEY

1.	B	21.	D	41.	D
2.	C	22.	B	42.	B
3.	A	23.	A	43.	D
4.	D	24.	C	44.	C
5.	C	25.	A	45.	A
6.	B	26.	B	46.	C
7.	C	27.	D	47.	B
8.	A	28.	C	48.	B
9.	D	29.	A	49.	A
10.	C	30.	B	50.	D
11.	A	31.	B		
12.	C	32.	C		
13.	B	33.	A		
14.	B	34.	D		
15.	C	35.	D		
16.	D	36.	D		
17.	A	37.	A		
18.	D	38.	D		
19.	C	39.	A		
20.	B	40.	B		

PREPARATION EXAM 6—EXPLANATIONS

TASK D.15

1. One injector driver has failed several times. What should be checked to find the cause of this repeated failure?

A. Injector relay voltage drop
B. Injector resistance
C. ECM battery voltage
D. Injector ground resistance

Answer A is incorrect. Excessive voltage drop of the relay would not damage the injector driver.

Answer B is correct. Injector resistance should be tested. If the injector is shorted, the excessive current flow could cause damage to the injector driver.

Answer C is incorrect. An ECM battery voltage fault may damage the ECM, but it also would affect more than just one injector driver circuit.

Answer D is incorrect. The injector ground should have very low resistance. High resistance would not damage the injector driver.

2. A rich air/fuel ratio will cause HO$_2$S voltage to be:

TASK B.8

   A. Below 100 mV
   B. Above 100 mV
   C. Above 450 mV
   D. Fixed at 450 mV

Answer A is incorrect. A HO$_2$S voltage below 100 mV is considered a lean mixture.

Answer B is incorrect. A HO$_2$S voltage above 100 mV and below 450 mV is considered a lean mixture.

Answer C is correct. A HO$_2$S voltage above 450 mV is considered a rich mixture.

Answer D is incorrect. A HO$_2$S voltage fixed at 450 mV is considered to be a transitional voltage and is not rich or lean.

3. Which of the following customer concerns might be present with a stuck closed exhaust gas recirculation (EGR) valve?

TASK E.8

   A. Detonation
   B. Engine stalling
   C. Rough idle
   D. Low power output

Answer A is correct. The EGR valve prevents detonation by cooling the combustion chamber.

Answer B is incorrect. A stuck open EGR valve may cause engine stalling.

Answer C is incorrect. A stuck open EGR valve may cause a rough idle.

Answer D is incorrect. While there are many things that may cause an engine to have low power output, a stuck closed EGR valve is not one of them.

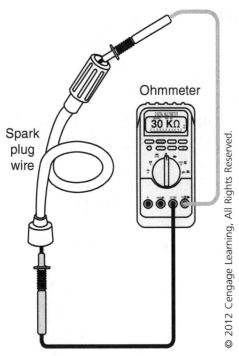

TASK C.5

4. The meter shown above displays 30 KΩ. Technician A says that the wire has too low resistance. Technician B says that the displayed reading would cause the firing line to be lower than normal when viewed on a scope. Who is correct?

   A. A only

   B. B only

   C. Both A and B

   D. Neither A nor B

   Answer A is incorrect. The meter is displaying a resistance of 30,000 Ω. This is excessive and the wire should be replaced.

   Answer B is incorrect. Due to the high resistance, this wire would cause the firing line to be much higher than normal when viewed on a scope.

   Answer C is incorrect. Neither Technician is correct.

   Answer D is correct. Neither Technician is correct. Spark plug wire resistance should be no more than 3000 to 7000 Ω per foot.

TASK A.14

5. A catalytic converter has failed several times with efficiency problems. Technician A says that this could be caused by an exhaust manifold leak. Technician B says that this could be caused by a seeping head gasket. Who is correct?

   A. A only

   B. B only

   C. Both A and B

   D. Neither A nor B

   Answer A is incorrect. Technician B is also correct.

   Answer B is incorrect. Technician A is also correct.

   Answer C is correct. Both Technicians are correct. An exhaust manifold leak can disrupt the activity of the oxygen sensor, which can alter the results of the catalyst monitor. A seeping head gasket that is allowing coolant to be burned can damage the catalytic converter if the condition is not corrected.

   Answer D is incorrect. Both Technicians are correct.

6. O_2 sensor voltage is fixed high. Which of the following emissions gases would be expected to increase?

    A. O_2
    B. CO
    C. NO_x
    D. CO_2

    TASK B.11

    Answer A is incorrect. High O_2 levels are likely to be present with lean air/fuel ratios.

    Answer B is correct. High O_2 sensor voltage indicates a rich condition. CO emissions are higher with rich air/fuel ratios.

    Answer C is incorrect. NO_x emissions are likely to be higher with lean air/fuel ratios.

    Answer D is incorrect. Air/fuel ratios other than 14.7:1 will lower CO_2 emissions.

7. Technician A says that a vehicle will be rejected from the I/M test if one or more of the monitors have not run. Technician B says that a vehicle will be rejected if the MIL does not work. Who is correct?

    A. A only
    B. B only
    C. Both A and B
    D. Neither A nor B

    TASK F.16

    Answer A is incorrect. Technician B is also correct.

    Answer B is incorrect. Technician A is also correct.

    Answer C is correct. Both Technicians are correct. A vehicle will be rejected from I/M testing if the MIL is inoperative or if one or more of the monitors have not run.

    Answer D is incorrect. Both Technicians are correct.

8. A vehicle has entered the shop with a DTC P0401 – insufficient EGR flow. Which of the following modifications can cause this condition?

    A. Installing a high-flow exhaust system
    B. Installing a high-flow air intake system
    C. Installing a larger throttle body
    D. Installing a larger MAF sensor

    TASK A.11

    Answer A is correct. The proper amount of backpressure is needed for correct EGR gas flow.

    Answer B is incorrect. A high-flow air intake system will not affect EGR system flow, but it is likely to affect MAF sensors and cold start emissions.

    Answer C is incorrect. Installing a larger throttle body will not affect EGR system flow.

    Answer D is incorrect. A larger MAF sensor will not affect EGR system flow.

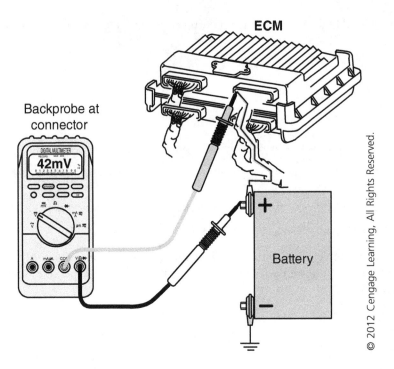

ECM

Backprobe at connector

42mV

Battery

© 2012 Cengage Learning, All Rights Reserved.

TASK B.14

9. The ECM power circuit is being tested as shown above. Technician A says that the circuit has high resistance since the meter is not displaying 12 V. Technician B says that corrosion in the switch can cause a reading that is higher than 0 V. Who is correct?

A. A only

B. B only

C. Both A and B

D. Neither A nor B

Answer A is incorrect. The meter is showing the voltage drop of the power circuit. This measurement should be low and is displaying an acceptable measurement.

Answer B is incorrect. If the switch had excessive corrosion, the meter would display a high voltage drop.

Answer C is incorrect. Neither Technician is correct.

Answer D is correct. Neither Technician is correct. Many circuits will have some voltage drop. Usually any measurement less than 0.2 V is acceptable.

TASK E.10

10. The AIR injection tube has rusted off of the vehicle's catalytic converter. Which two emissions will be increased by this fault?

A. HC and CO_2

B. CO and NO_x

C. HC and CO

D. O_2 and CO_2

Answer A is incorrect. The described failure will increase HC but not CO_2.

Answer B is incorrect. The described failure will increase CO but not NO_x.

Answer C is correct. The purpose of injecting oxygen into the catalytic converter is to aid in the oxidation of HC and CO. A lack of oxygen will increase these two gases.

Answer D is incorrect. The described failure will decrease O_2 or CO_2.

11. A vehicle equipped with a distributor ignition (DI) exhibits spark knock. Which of the following conditions could cause this?

TASK C.9

    A. Too-advanced ignition timing

    B. Too-retarded ignition timing

    C. Switched plug wires

    D. High resistance in the plug wires

Answer A is correct. Ignition timing that is too far advanced can lead to a spark knock condition.

Answer B is incorrect. Ignition timing that is too retarded will not cause spark knock, but rather a lack of power instead.

Answer C is incorrect. Switched plug wires will cause an engine to misfire.

Answer D is incorrect. High resistance in the plug wires will cause the engine to misfire under heavy load.

12. A vehicle has been brought to the shop with a concern of lowered fuel economy. Technician A says that the use of oxygenated fuels such as RFG (reformulated gasoline) can cause lowered fuel economy. Technician B says that the use of fuels with high alcohol content can cause lowered fuel economy. Who is correct?

TASK D.11

    A. A only

    B. B only

    C. Both A and B

    D. Neither A nor B

Answer A is incorrect. Technician B is also correct.

Answer B is incorrect. Technician A is also correct.

Answer C is correct. Both Technicians are correct. RFG has a lower energy potential than regular gasoline; therefore, a vehicle will get lower fuel economy when using this fuel. Alcohol based fuels have a lower energy potential than gasoline, so when the alcohol content of a fuel increases, the energy content decreases.

Answer D is incorrect. Both Technicians are correct.

13. A vehicle starts and immediately dies. Which of the following is the most likely cause of this condition?

TASK B.13

    A. TPS voltage above 4.0 V

    B. An immobilizer fault

    C. Low fuel pressure

    D. Contaminated MAF sensor

Answer A is incorrect. TPS voltage above 4.0 V will cause the ECM to enter clear flood mode. This will not allow the vehicle to start at all.

Answer B is correct. An immobilizer system fault will cause the vehicle to start and immediately die.

Answer C is incorrect. Low fuel pressure will cause the vehicle to have a lack of power or prevent it from starting at all if excessively low.

Answer D is incorrect. A contaminated MAF sensor will not cause the vehicle to start and die, but rather set a DTC and cause the MIL to illuminate.

TASK F.10

14. A vehicle has failed a loaded I/M test for excessive HC emissions and a misfire. Technician A says that a restricted fuel filter could cause this condition. Technician B says that a fault in the secondary ignition system could cause this condition. Who is correct?

   A. A only

   B. B only

   C. Both A and B

   D. Neither A nor B

   Answer A is incorrect. A restricted fuel filter would cause a lean condition under load, which would elevate O_2 emissions and lower CO_2 levels.

   Answer B is correct. Only Technician B is correct. A fault in the secondary ignition system could cause the vehicle to miss under load and have high HC emissions.

   Answer C is incorrect. Only Technician B is correct.

   Answer D is incorrect. Technician B is correct.

TASK A.8

15. A vehicle has failed the I/M emissions test for excessive NO_x output. Which of the following conditions is most likely to cause elevated NO_x emissions?

   A. Retarded ignition timing

   B. A high-flow air intake system

   C. Debris-plugged radiator

   D. Excessive fuel pressure

   Answer A is incorrect. Retarded ignition timing may cause increased HCs and CO but will not increase NO_x levels.

   Answer B is incorrect. A high-flow air intake system is not likely to cause increased NO_x emissions but is likely to cause increased O_2 emissions.

   Answer C is correct. Lack of airflow through the radiator can cause elevated engine temperatures, which will increase NO_x emissions.

   Answer D is incorrect. Excessive fuel pressure will cause a rich condition. A lean condition is more likely to cause excessive NO_x emissions.

TASK C.9

16. An engine with a DI has no spark and will not start. A test light is connected to the negative coil circuit and flashes while the vehicle is being cranked. Which of the following components could be faulty?

   A. Crank sensor

   B. Ignition module

   C. Ignition fuse

   D. Ignition coil

   Answer A is incorrect. The primary circuit is functioning properly, since the test lamp is flashing.

   Answer B is incorrect. The test lamp would not flash if the ignition module were not functioning.

   Answer C is incorrect. The test lamp would not flash if the ignition fuse were blown.

   Answer D is correct. The ignition coil could have a problem that is not allowing it to spark.

17. The HO$_2$S heater circuit is being tested with an ohmmeter. What would a reading of O.L. indicate?

    TASK B.16

    A. The circuit is open
    B. The circuit is shorted
    C. There is excessive corrosion
    D. The circuit is functional

    Answer A is correct. An ohmmeter reading of O.L. indicates that the circuit is open.

    Answer B is incorrect. A shorted circuit would have a measurement that is lower than specification.

    Answer C is incorrect. Excessive corrosion would cause a resistance measurement that is higher than specification.

    Answer D is incorrect. An ohmmeter reading of O.L. indicates that the circuit is open.

18. Which of the following would cause an increase in CO emissions?

    TASK F.5

    A. An intake manifold gasket leak
    B. Clogged injectors
    C. A contaminated MAF sensor
    D. Excessive fuel pressure

    Answer A is incorrect. CO is a rich indicator, and a leaking intake manifold gasket would cause a lean condition.

    Answer B is incorrect. CO is a rich indicator, and clogged injectors would cause a lean condition.

    Answer C is incorrect. A contaminated MAF sensor would cause unmetered air to enter the engine, which would cause a lean condition.

    Answer D is correct. Excessive fuel pressure would cause a rich condition and increase CO emissions.

19. A vehicle has entered the shop several times with a plugged fuel filter and now has entered the shop with a faulty fuel pump. Technician A says that the tank should be cleaned before assembly. Technician B says that the customer should find a different fuel supplier. Who is correct?

    TASK D.14

    A. A only
    B. B only
    C. Both A and B
    D. Neither A nor B

    Answer A is incorrect. Technician B is also correct.

    Answer B is incorrect. Technician A is also correct.

    Answer C is correct. Both Technicians are correct. The vehicle shows symptoms of having a lot of debris in the fuel. In this case, the tank should be removed and cleaned. The customer should also consider purchasing fuel at a different outlet to prevent this problem in the future.

    Answer D is incorrect. Both Technicians are correct.

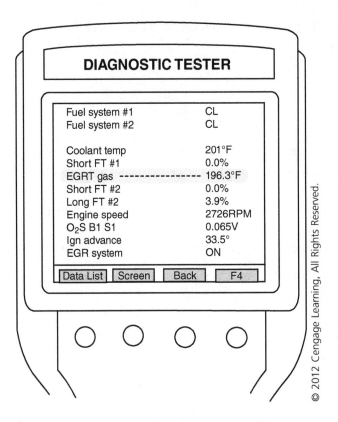

TASK E.9

20. While the EGR valve is being commanded open, the scan tool data in the above figure is displayed. Technician A says that the EGR temperature sensor is faulty. Technician B says that this condition will illuminate the MIL. Who is correct?

   A. A only

   B. B only

   C. Both A and B

   D. Neither A nor B

Answer A is incorrect. Since the EGR temperature sensor is reading fairly close to the coolant temperature sensor, it should not be considered faulty.

Answer B is correct. Only Technician B is correct. The EGR temperature sensor detects the heat from the exhaust gas that enters the intake. If the temperature does not rise when the EGR valve is opened, a DTC, P0401 – EGR insufficient flow will cause the MIL to illuminate.

Answer C is incorrect. Only Technician B is correct.

Answer D is incorrect. Technician B is correct.

21. The ECT sensor is being tested for operation. Technician A says that sensor resistance should increase as its temperature increases. Technician B says that the sensor voltage should increase as its temperature increases. Who is correct?

    A. A only

    B. B only

    C. Both A and B

    D. Neither A nor B

    TASK B.17

    Answer A is incorrect. The resistance of the ECT sensor should decrease as it is heated.

    Answer B is incorrect. The sensor voltage should decrease as it is heated.

    Answer C is incorrect. Neither Technician is correct.

    Answer D is correct. Neither Technician is correct. Most temperature sensors used on vehicles are negative temperature coefficient thermistors. The resistance and voltage signal of these sensors decrease as the temperature increases.

22. What effect can low-quality air filters have on MAF equipped vehicles?

    A. Cause a rich condition

    B. Increase NO$_x$ emissions

    C. Cause a low idle speed

    D. Cause a rough idle

    TASK A.11

    Answer A is incorrect. MAF contamination is likely to cause a lean condition due to the unmetered air entering the engine.

    Answer B is correct. Low-quality air filters can lead to MAF sensor contamination and a lean condition. Lean mixtures tend to burn hotter, which can result in increased NO$_x$ emissions.

    Answer C is incorrect. Low-quality air filters will not have an effect on idle speed.

    Answer D is incorrect. Low-quality air filters will not cause a rough idle.

23. During routine maintenance, a technician discovers that the air filter is soaked with oil. Which of the following would be the most likely cause of this problem?

    A. A plugged PCV valve

    B. An overfilled crankcase

    C. A plugged PCV vent

    D. Excessive oil pressure

    TASK E.8

    Answer A is correct. A plugged PCV valve would cause crankcase pressure to get high. The pressure would escape through the PCV vent hose and cause the air filter to become contaminated with oil.

    Answer B is incorrect. An overfilled crankcase would not cause the air filter to become oil soaked, but it may cause low or erratic oil pressures.

    Answer C is incorrect. A plugged PCV vent would not allow fresh air to enter the crankcase, which would not allow oil to reach the air filter.

    Answer D is incorrect. Excessive oil pressure would not cause oil to reach the air filter; however, engine damage could still result.

TASK B.14

24. The voltage drop of the ECM ground circuit is 0.8 V. What does this indicate?

 A. A short in the circuit

 B. An open in the circuit

 C. High resistance in the circuit

 D. Low resistance in the circuit

Answer A is incorrect. A voltage drop test cannot be used to measure for shorts in most circuits.

Answer B is incorrect. An open in the circuit would cause the voltage drop of the circuit to be source voltage.

Answer C is correct. A high voltage drop in a circuit indicates excessive resistance.

Answer D is incorrect. Voltage drop testing cannot be used to determine if a circuit has lower than specified resistance.

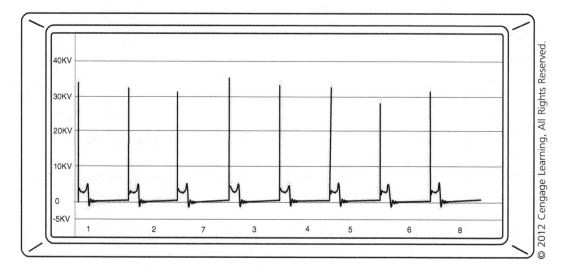

TASK C.10

25. Which of the following could cause the secondary ignition waveforms shown above?

 A. Faulty coil wire

 B. Faulty plug wire

 C. Fouled spark plug

 D. Rich air/fuel mixture

Answer A is correct. A coil wire with high resistance would cause all firing voltages to be higher than normal.

Answer B is incorrect. A faulty plug wire would cause the firing line of only one cylinder to be higher than the others.

Answer C is incorrect. A fouled spark plug would cause a low firing line for only one cylinder.

Answer D is incorrect. A rich air/fuel ratio will cause firing voltages to decrease and spark lines that slant downward.

26. An SFI vehicle will not start. Neither spark nor injector pulse is present. Technician A says that a TPS voltage fixed above 4.0 V will cause this condition. Technician B says that a faulty CKP sensor can cause this condition. Who is correct?

TASK B.11

    A. A only

    B. B only

    C. Both A and B

    D. Neither A nor B

    Answer A is incorrect. A TPS voltage fixed above 4.0 V will cause the vehicle to enter clear flood mode. This will cause the vehicle to have no injector pulse, but the ECM will not disable spark.

    Answer B is correct. Only Technician B is correct. A faulty CKP sensor will prevent the ECM from receiving any RPM data, and it will not be able to calculate ignition or fuel injector timing.

    Answer C is incorrect. Only Technician B is correct.

    Answer D is incorrect. Technician B is correct.

27. Technician A says that high CO emissions are a good indicator of a lean condition. Technician B says that high O_2 levels are a good indicator of a rich condition. Who is correct?

TASK F.4

    A. A only

    B. B only

    C. Both A and B

    D. Neither A nor B

    Answer A is incorrect. High CO levels indicate a rich condition.

    Answer B is incorrect. High O_2 levels indicate a lean condition.

    Answer C is incorrect. Neither Technician is correct.

    Answer D is correct. Neither Technician is correct. CO in the exhaust is a rich indicator, while O_2 in the exhaust is a lean indicator.

28. An exhaust manifold leak can cause which of the following:

TASK A.7

    A. High O_2 sensor readings

    B. Negative fuel trim numbers

    C. Low O_2 sensor readings

    D. Low MAF sensor readings

    Answer A is incorrect. High O_2 sensor voltages indicate a lack of oxygen in the exhaust. An exhaust leak will increase the oxygen content of the exhaust.

    Answer B is incorrect. An exhaust manifold leak would allow excessive oxygen in the exhaust, which will cause a lean O_2 sensor reading and positive fuel trim numbers.

    Answer C is correct. An exhaust manifold leak can cause excess oxygen to enter the exhaust stream and be picked up by the O_2 sensor. The sensor will display a low voltage.

    Answer D is incorrect. An exhaust manifold leak will not affect the readings of the MAF sensor since they are in different systems.

TASK F.5

29. A vehicle has a rough and unstable idle. A check of the exhaust shows elevated HCs, CO_2 at 11 percent, O_2 at six percent, and CO less than one percent. Which of the following is most likely to cause these conditions?

    A. A vacuum leak

    B. A leaking injector

    C. A dirty throttle body

    D. A faulty fuel pump check valve

    Answer A is correct. The elevated O_2 and HC levels can indicate a lean misfire, which could be caused by a vacuum leak.

    Answer B is incorrect. A leaking injector would cause a rich condition with elevated CO levels.

    Answer C is incorrect. A dirty throttle body could cause a slightly rich condition.

    Answer D is incorrect. A faulty fuel pump check valve would cause hard starting and would not affect emissions while the vehicle is running.

TASK B.15

30. The fuel injectors are being current tested on a vehicle, and one is found to be high. Which of the following can cause this?

    A. High resistance

    B. Low resistance

    C. Low voltage

    D. High fuel pressure

    Answer A is incorrect. High resistance will decrease current flow.

    Answer B is correct. Low resistance will increase current flow.

    Answer C is incorrect. Low voltage will decrease current flow.

    Answer D is incorrect. High fuel pressure will not affect the current flow of the injector.

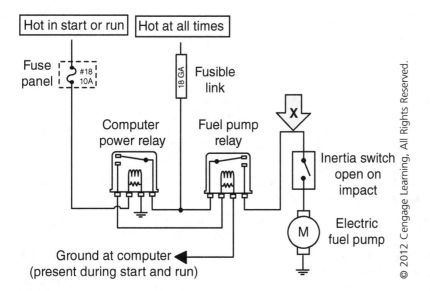

TASK D.8

31. A vehicle is brought to the shop for a no-start concern, and it has been determined that the fuel pump is not running. In the above figure, the voltage at point X in the circuit, measured while the fuel pump is commanded on, is 0 V. Which of the following could cause this problem?

   A. A stuck closed fuel pump relay
   B. An open fuel pump relay coil
   C. An open fuel pump ground
   D. A tripped inertia switch

   Answer A is incorrect. A stuck closed fuel pump relay would cause the fuel pump to run at all times.

   Answer B is correct. An open fuel pump relay coil will not allow the relay to operate and will prevent the fuel pump from working.

   Answer C is incorrect. If the fuel pump ground were open, there would still be voltage at point X.

   Answer D is incorrect. A tripped inertia switch would open the circuit after point X, and voltage would be measured.

TASK B.5

32. Technician A says that during open loop, the vehicle's fuel needs are calculated using engine sensors such as the MAF, TPS, and ECT. Technician B says that during closed loop, fuel calculations are made based on O_2 sensor readings. Who is correct?

   A. A only
   B. B only
   C. Both A and B
   D. Neither A nor B

   Answer A is incorrect. Technician B is also correct.

   Answer B is incorrect. Technician A is also correct.

   Answer C is correct. Both Technicians are correct. Fuel calculations in open loop are made by readings from sensors such as the MAF, TPS, and ECT. Fuel calculations are made based on O_2 sensor readings while in closed loop.

   Answer D is incorrect. Both Technicians are correct.

TASK D.5

33. Negative fuel trim values can be caused by all of the following EXCEPT:

　　A. A contaminated MAF sensor

　　B. Excessive fuel pressure

　　C. A leaking fuel injector

　　D. A leaking EVAP purge valve

Answer A is correct. A contaminated MAF sensor will cause unmetered air to enter the engine. This unmetered air will cause a lean condition, which will cause fuel trim numbers to increase.

Answer B is incorrect. Excessive fuel pressure will cause a rich condition and negative fuel trim numbers due to the excessive amount of fuel entering the engine.

Answer C is incorrect. A leaking fuel injector will cause a rich condition and negative fuel trim numbers, since it will cause one cylinder to receive too much fuel.

Answer D is incorrect. A leaking EVAP purge valve will cause fuel vapors to enter the intake at all times and cause negative fuel trim numbers.

TASK F.13

34. Which of the following would LEAST LIKELY cause a vehicle to immediately fail the I/M test?

　　A. No MIL illumination

　　B. An illuminated MIL

　　C. A missing fuel cap

　　D. Low fuel level

Answer A is incorrect. An inoperative MIL will cause a vehicle to immediately fail the I/M test if it does not turn on during the bulb test.

Answer B is incorrect. A fully illuminated MIL will cause a vehicle to immediately fail an I/M test, since it is an indicator that the vehicle's emissions are already exceeding 1.5 times the federal standard.

Answer C is incorrect. A missing fuel cap will cause excessive evaporative emissions and immediate failure of the I/M test.

Answer D is correct. Low fuel level will not affect the I/M test.

TASK F.9

35. A vehicle has failed the I/M test for excessive CO at idle. Which of the following problems could cause this condition?

　　A. A manifold vacuum leak

　　B. A restricted fuel filter

　　C. A restricted catalytic converter

　　D. A faulty AIR pump

Answer A is incorrect. A manifold vacuum leak would cause a lean condition and low CO emissions.

Answer B is incorrect. A restricted fuel filter would cause a lean condition and low CO emissions.

Answer C is incorrect. A malfunctioning catalytic converter could cause excessive CO emissions, but a restricted catalytic converter would not cause excessive CO emissions at idle.

Answer D is correct. The AIR adds oxygen to the catalytic converter to aid in converting CO to CO_2.

36. A catalytic converter is suspected faulty. Which of the following emissions would be LEAST LIKELY to increase?

TASK E.10

   A. HC

   B. NO_x

   C. CO

   D. CO_2

   Answer A is incorrect. HC levels would be expected to increase if the catalytic converter were not oxidizing it to form H_2O and CO_2.

   Answer B is incorrect. NO_x levels would be expected to increase if the catalytic converter were not reducing it to N_2 and O_2.

   Answer C is incorrect. CO levels would be expected to increase if the catalytic converter were not oxidizing it into CO_2.

   Answer D is correct. CO_2 is highest when the engine is running efficiently and the catalytic converter is operating efficiently. A faulty catalytic converter would cause CO_2 levels to decrease.

37. The composite vehicle has entered the shop with a DTC P0011 – camshaft over advanced bank 1. Technician A says that this could be caused by debris stuck in the camshaft position actuator control solenoid. Technician B says that this could be caused by low oil pressure. Who is correct?

TASK A.7

   A. A only

   B. B only

   C. Both A and B

   D. Neither A nor B

   Answer A is correct. Only Technician A is correct. Debris stuck in the camshaft position actuator control solenoid will cause the solenoid to be stuck open. This will cause oil pressure to advance the camshaft to the fully advanced position.

   Answer B is incorrect. Low oil pressure will cause the camshaft position actuator to default to the fully retarded position.

   Answer C is incorrect. Only Technician A is correct.

   Answer D is incorrect. Technician A is correct.

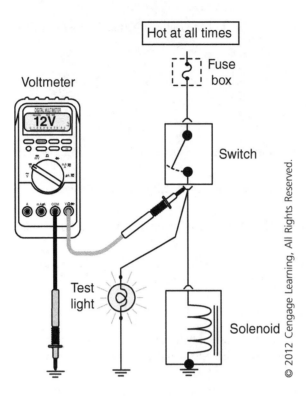

TASK B.5

38. The solenoid in the circuit is inoperative. While testing with the meter shown in the above figure, the circuit displays 12 V, and the test light illuminates when the switch is closed. All of the following can cause this EXCEPT:

A. An open ground circuit

B. An open solenoid

C. An open wire between the switch and the solenoid

D. An open switch

Answer A is incorrect. An open ground circuit would cause the solenoid not to function and voltage after the switch.

Answer B is incorrect. An open solenoid would cause the circuit not to work and voltage after the switch.

Answer C is incorrect. An open in the wire from the switch to the solenoid would cause the solenoid not to function and voltage after the switch.

Answer D is correct. An open switch would cause the meter to display 0 V and prevent the test light from illuminating.

39. A late-model vehicle with on-board refueling vapor recovery (ORVR) is slow to refuel. Which of the following could cause this problem?

    A. A restricted vent filter

    B. A stuck open vent solenoid

    C. A stuck closed purge solenoid

    D. A restricted purge line

TASK E.8

    Answer A is correct. While refueling, the air that is in the tank exits through the vent solenoid. Many of these systems will have a filter to clean incoming air. If this filter becomes restricted the tank will be slow to fill.

    Answer B is incorrect. The vent solenoid is normally open and will not cause this problem.

    Answer C is incorrect. The purge solenoid is normally closed and will not cause this problem.

    Answer D is incorrect. Air does not pass through the purge line when refueling and will not cause the tank to be slow to fill if restricted.

40. A vehicle failed the I/M test for excessive CO emissions and its O_2 sensor readings are fixed above 850 mV. All of the following could cause this EXCEPT:

    A. ECT sensor

    B. Skewed O_2 sensor

    C. Leaky fuel pressure regulator

    D. A skewed MAP sensor

TASK B.8

    Answer A is incorrect. A faulty ECT sensor can cause a rich condition.

    Answer B is correct. High CO emissions and high O_2 sensor voltage both indicate a rich condition. Since both of these indicate the same thing, the O_2 sensor should not be considered faulty.

    Answer C is incorrect. A leaking fuel pressure sensor can cause a rich condition.

    Answer D is incorrect. A skewed MAP sensor can cause a rich condition.

41. A catalytic converter can be damaged by all of the following EXCEPT:

    A. A faulty O_2 sensor

    B. A blown head gasket

    C. A faulty plug wire

    D. A dirty air filter

TASK E.13

    Answer A is incorrect. A faulty O_2 sensor will damage the catalytic converter if it continually sends a lean signal. The ECM will command a rich mixture, which will damage the converter.

    Answer B is incorrect. A blown head gasket will damage the catalytic converter due to coolant and excessive HCs entering the converter.

    Answer C is incorrect. A faulty plug wire will damage the catalytic converter by allowing excessive unburned HCs to enter the converter.

    Answer D is correct. While a dirty air filter could cause a rich condition, the ECM can adapt to these conditions and lean out the mixture to prevent the converter from being damaged.

TASK B.13

42. Intermittently, the composite vehicle will not start, and the anti-theft indicator lamp will flash. Technician A says that the battery in the key may be getting low. Technician B says that key rings containing other transponder keys for different vehicles can affect the signal received by the antenna. Who is correct?

   A. A only
   B. B only
   C. Both A and B
   D. Neither A nor B

Answer A is incorrect. Transponder keys are manufactured to last the lifetime of the vehicle.

Answer B is correct. Only Technician B is correct. When different vehicles' transponder keys are close to the antenna, they can interfere with the correct signal and cause intermittent no-start problems.

Answer C is incorrect. Only Technician B is correct.

Answer D is incorrect. Technician B is correct.

TASK C.3

43. The composite vehicle is brought into the shop with a miss on cylinder 1 and a suspected faulty ignition coil. Technician A says that the resistance from coil terminal "A" to terminal "B" should be 10 KΩ. Technician B says that the resistance from terminal "A" to terminal "C" should be 1 Ω. Who is correct?

   A. A only
   B. B only
   C. Both A and B
   D. Neither A nor B

Answer A is incorrect. Terminals "A" and "B" are for the primary coil, which should have a resistance of 1 Ω.

Answer B is incorrect. Terminals "A" and "C" are for the secondary ignition coil and should have a resistance of 10 KΩ.

Answer C is incorrect. Neither Technician is correct.

Answer D is correct. Neither Technician is correct. The ignition coil for the composite vehicle should have a primary resistance of 1 Ω and a secondary resistance of 10 KΩ.

TASK B.14

44. The brake pedal position switch of the composite vehicle is being tested. The voltage drop across the switch with the pedal depressed is 12.6 V. This voltage indicates:

   A. Low switch input voltage
   B. A closed brake switch
   C. An open brake switch
   D. High power circuit resistance

Answer A is incorrect. If the switch were functioning properly, the voltage drop would be 0 V, even if input voltage were low.

Answer B is incorrect. A closed switch would have a voltage drop of 0 V.

Answer C is correct. When the brake pedal is depressed, the switch should be closed. The voltage drop across the switch in this position should be close to 0 V. A 12 V drop indicates an open switch.

Answer D is incorrect. The voltage drop across the switch would still be 0 V if the power circuit had excessive resistance. A switch should not use any voltage, since it is not an electrical load.

45. The composite vehicle will enter closed loop when:

    A. Valid AFR sensor readings are received by the ECM

    B. Coolant temperature reaches 195°F (91°C)

    C. The engine has run for at least ten minutes

    D. TPS is less than 50 percent

TASK D.3

Answer A is correct. Once both upstream O_2 sensors have heated up enough to produce varying signals, the ECM can begin to use data from those sensors to calculate fuel delivery.

Answer B is incorrect. There are no coolant temperature minimums for closed loop operation.

Answer C is incorrect. There are no time restrictions for closed loop operation.

Answer D is incorrect. Closed loop can occur with the TPS less than 80 percent.

46. The composite vehicle will not start, and the CKP sensor is suspected faulty. To check the sensor's output while cranking, what setting should the DVOM be set to?

    A. AC amps

    B. DC amps

    C. AC volts

    D. DC volts

TASK C.4

Answer A is incorrect. The CKP sensor's amperage cannot be measured.

Answer B is incorrect. The CKP sensor's amperage cannot be measured.

Answer C is correct. The composite vehicle's CKP sensor is a magnetic generator type sensor that produces an AC voltage.

Answer D is incorrect. The CKP sensor produces AC voltage.

47. Which of the following would be an acceptable resistance measurement of the composite vehicle's fuel injector?

    A. 5 Ω

    B. 10 Ω

    C. 10 KΩ

    D. O.L.

TASK D.3

Answer A is incorrect. 5 Ω would be too low for the fuel injector resistance. This would indicate a shorted injector.

Answer B is correct. The resistance specification for the composite vehicle's fuel injector is 12 ± 2 Ω.

Answer C is incorrect. 10 KΩ would be too high of an injector resistance. An injector with resistance this high should be replaced.

Answer D is incorrect. A resistance reading of O.L. would indicate an open injector.

TASK E.9

48. The composite vehicle has entered the shop with a DTC P0440 – EVAP system leak. Technician A says that the system can be leak tested with a smoke machine connected to shop air. Technician B says that the vent solenoid must be commanded shut. Who is correct?

    A. A only

    B. B only

    C. Both A and B

    D. Neither A nor B

Answer A is incorrect. While a smoke machine is used to check the EVAP system for leaks, it should never be connected to shop air to pressurize the system. An inert gas, such as nitrogen, should be used to pressurize the EVAP system.

Answer B is correct. Only Technician B is correct. To leak test the EVAP system, the vent solenoid should be commanded shut to seal the system.

Answer C is incorrect. Only Technician B is correct.

Answer D is incorrect. Technician B is correct.

49. The composite vehicle has failed the I/M test for excessive NO_x emissions. Which of the following could cause this condition?

    A. An open at ECM terminal 108

    B. An open at ECM terminal 110

    C. An open at ECM terminal 160

    D. An open at ECM terminal 240

TASK F.3

Answer A is correct. ECM terminal 108 controls the EGR valve. An open here would prevent the EGR system from functioning and would increase NO_x emissions.

Answer B is incorrect. ECM terminal 110 controls the EVAP purge solenoid, which has no effect on NO_x emissions. The EVAP solenoid would not function if this terminal were open, and this would cause an EVAP system failure.

Answer C is incorrect. ECM terminal 160 is the downstream HO_2S signal wire, which will have no effect on NO_x emissions. An open on this terminal would prevent the function of the downstream O_2 sensor.

Answer D is incorrect. ECM terminal 240 connects to the bank 1 knock sensor and will have no effect on NOx emissions.

TASK E.9

50. The composite vehicle's EVAP system is being tested. When the engine is running and the purge and vent solenoids are commanded on, the fuel tank pressure sensor voltage steadily decreases. This drop in voltage indicates:

    A. Insufficient vacuum supply

    B. A stuck open vent valve

    C. A faulty fuel tank pressure sensor

    D. A sealed EVAP system

Answer A is incorrect. Insufficient vacuum supply to the EVAP system would prevent fuel tank pressure sensor voltage from decreasing.

Answer B is incorrect. Fuel tank pressure sensor voltage would not decrease at all if the vent valve were stuck open.

Answer C is incorrect. The fuel tank pressure sensor is working as designed and should not be considered faulty.

Answer D is correct. With the purge and vent valves commanded on, the vent should be closed and vacuum applied to the system. As the pressure in the tank decreases, the fuel tank pressure sensor voltage should also decrease.

Appendices

PREPARATION EXAM ANSWER SHEET FORMS

ANSWER SHEET

1. _____	21. _____	41. _____
2. _____	22. _____	42. _____
3. _____	23. _____	43. _____
4. _____	24. _____	44. _____
5. _____	25. _____	45. _____
6. _____	26. _____	46. _____
7. _____	27. _____	47. _____
8. _____	28. _____	48. _____
9. _____	29. _____	49. _____
10. _____	30. _____	50. _____
11. _____	31. _____	
12. _____	32. _____	
13. _____	33. _____	
14. _____	34. _____	
15. _____	35. _____	
16. _____	36. _____	
17. _____	37. _____	
18. _____	38. _____	
19. _____	39. _____	
20. _____	40. _____	

ANSWER SHEET

1. _____	21. _____	41. _____
2. _____	22. _____	42. _____
3. _____	23. _____	43. _____
4. _____	24. _____	44. _____
5. _____	25. _____	45. _____
6. _____	26. _____	46. _____
7. _____	27. _____	47. _____
8. _____	28. _____	48. _____
9. _____	29. _____	49. _____
10. _____	30. _____	50. _____
11. _____	31. _____	
12. _____	32. _____	
13. _____	33. _____	
14. _____	34. _____	
15. _____	35. _____	
16. _____	36. _____	
17. _____	37. _____	
18. _____	38. _____	
19. _____	39. _____	
20. _____	40. _____	

ANSWER SHEET

1. _____	21. _____	41. _____
2. _____	22. _____	42. _____
3. _____	23. _____	43. _____
4. _____	24. _____	44. _____
5. _____	25. _____	45. _____
6. _____	26. _____	46. _____
7. _____	27. _____	47. _____
8. _____	28. _____	48. _____
9. _____	29. _____	49. _____
10. _____	30. _____	50. _____
11. _____	31. _____	
12. _____	32. _____	
13. _____	33. _____	
14. _____	34. _____	
15. _____	35. _____	
16. _____	36. _____	
17. _____	37. _____	
18. _____	38. _____	
19. _____	39. _____	
20. _____	40. _____	

ANSWER SHEET

1. _____	21. _____	41. _____
2. _____	22. _____	42. _____
3. _____	23. _____	43. _____
4. _____	24. _____	44. _____
5. _____	25. _____	45. _____
6. _____	26. _____	46. _____
7. _____	27. _____	47. _____
8. _____	28. _____	48. _____
9. _____	29. _____	49. _____
10. _____	30. _____	50. _____
11. _____	31. _____	
12. _____	32. _____	
13. _____	33. _____	
14. _____	34. _____	
15. _____	35. _____	
16. _____	36. _____	
17. _____	37. _____	
18. _____	38. _____	
19. _____	39. _____	
20. _____	40. _____	

ANSWER SHEET

1. _____	21. _____	41. _____
2. _____	22. _____	42. _____
3. _____	23. _____	43. _____
4. _____	24. _____	44. _____
5. _____	25. _____	45. _____
6. _____	26. _____	46. _____
7. _____	27. _____	47. _____
8. _____	28. _____	48. _____
9. _____	29. _____	49. _____
10. _____	30. _____	50. _____
11. _____	31. _____	
12. _____	32. _____	
13. _____	33. _____	
14. _____	34. _____	
15. _____	35. _____	
16. _____	36. _____	
17. _____	37. _____	
18. _____	38. _____	
19. _____	39. _____	
20. _____	40. _____	

ANSWER SHEET

1. _____	21. _____	41. _____
2. _____	22. _____	42. _____
3. _____	23. _____	43. _____
4. _____	24. _____	44. _____
5. _____	25. _____	45. _____
6. _____	26. _____	46. _____
7. _____	27. _____	47. _____
8. _____	28. _____	48. _____
9. _____	29. _____	49. _____
10. _____	30. _____	50. _____
11. _____	31. _____	
12. _____	32. _____	
13. _____	33. _____	
14. _____	34. _____	
15. _____	35. _____	
16. _____	36. _____	
17. _____	37. _____	
18. _____	38. _____	
19. _____	39. _____	
20. _____	40. _____	

Glossary

The reference materials and questions for the L1 test use electronic and emission terms and acronyms that are consistent with the industry-wide SAE standards J1930 and J2012. Some of these terms are listed below.

Acceleration Simulation Mode (ASM) Loaded-mode steady-state tests that measure HC, CO and NOx emissions while the vehicle is driven on a dynamometer at a fixed speed and load. ASM5015 is a test at 15 mph with a load equivalent to 50 percent of the power needed to accelerate the vehicle at 3.3 mph per second. ASM2525 is a test at 25 mph with a load of 25 percent of the same power.

Air Injection Reaction (AIR) System Pumps oxygen into the exhaust to allow for the oxidation of unburned hydrocarbons (HCs).

Calculated Load Valve The percentage of engine capacity being used, based on current airflow divided by maximum airflow.

Data Link Connector (DLC) The standardized plug that is used to connect the scan tool to the computer.

Diagnostic Trouble Codes (DTC) Codes stored by the computer when a problem is detected and read using a scan tool. Each code corresponds to a particular problem. When a DTC is referred to in an L1 test question, the number and description will both be given. For instance, P0114=Intake Air Temperature Circuit Intermittent.

Distributor Ignition (DI) An ignition system that uses a distributor.

Early Fuel Evaporation (EFE) System This system is really old and not used today.

Electronic Control Module (ECM) Computer that is in charge of controlling engine operations. This module monitors engine sensors to determine the necessary output action and perform numerous tests.

Electronic Ignition (EI) An ignition system that has coils dedicated to specific spark plugs and does not use a distributor; often referred to as a distributorless ignition.

EVAP System An abbreviated term for the evaporative emissions system. This system stores evaporated hydrocarbons (HCs) from the fuel tank so that they may be burned in the engine at a later time.

Exhaust Gas Recirculation (EGR) System Directs exhaust gas back into the intake of the engine. The exhaust gas lowers combustion pressure and temperature to prevent the formation of NOx gases. The possibility of detonation is also lowered by this system because of the lower combustion temperatures.

Federal Test Procedure (FTP) A transient-speed mass emissions test conducted on a loaded dynamometer. This is the test that, by law, car manufacturers use to certify that new vehicles are in compliance for hydrocarbon, carbon monoxide, and oxides of nitrogen emissions; all vehicle models must pass the FTP before they may be sold in the United States.

Freeze-frame Operating conditions that are stored in the memory of the PCM at the instant a diagnostic trouble code is set.

Fuel Trim (FT) Fuel delivery adjustments based on closed-loop feedback. Values above the central value (0 percent) indicate increased injector pulse width. Values below the central value (0 pecent), indicate decreased injector pulse width. *Short term fuel trim* is based on rapidly switching oxygen sensor values. *Long term fuel trim* is a learned value used to compensate for continual deviation of the short term fuel trim from its central value.

Generator J1930 term for alternator (generating device that uses a diode rectifier).

I/M Tests Inspection and maintenance tests; vehicle emissions tests required by state governments. Some common types of I/M tests include no-load, ASM, and I/M240.

I/M240 A loaded-mode transient test that measures HC, CO, CO_2, and NOx emissions in grams/mile second by second, while the tested vehicle is driven at various speeds and loads on a dynamometer for 240 seconds. Another transient load test is the **BAR31**, a 31-second test cycle that includes an acceleration ramp similar to the IM240.

Malfunction Indicator Lamp (MIL) A lamp on the instrument panel that lights when the PCM detects an emission-related problem, similar to a "check engine" light.

Manifold Absolute Pressure (MAP) The pressure in the intake manifold reduced to a perfect vacuum. Because manifold vacuum is the difference between manifold absolute pressure and atmospheric pressure, all the vacuum readings in the composite vehicle are taken at sea level, where standard atmospheric pressure equals 101 kPa or 29.92 in. Hg.

Mass Air Flow (MAF) System A fuel-injection system uses a MAF sensor to measure the mass (weight) of the air drawn into the intake manifold, measured in grams per second.

No-load Tests that measure HC emissions in parts per million (ppm) and CO emissions in percent, while the vehicle is in neutral. Examples are idle and two-speed.

On-board Diagnostics (OBD) A diagnostic system contained in the PCM that monitors computer inputs and outputs for failures. OBD-II is an industry-standard, second-generation OBD system that monitors emission control systems for degradation as well as failures.

On-board Refueling Vapor Recovery (ORVR) An evaporative emissions (EVAP) system that prevents the escape of HC vapors to the atmosphere by directing fuel tank vapors to the EVAP charcoal canister during fueling.

Positive Crankcase Ventilation (PCV) System Directs unburned hydrocarbons (HCs) from the crankcase into the intake manifold so that they can be burned in the cylinder.

Pulse Width Modulation (PWM) An electronic signal with a variable on-off time.

Powertrain Control Module (PCM) The electronic computer that controls the engine and transmission; similar to an ECM.

Root Cause of Failure A component or system failure which, if not repaired, can cause other failures. If the secondary failure is repaired but the root cause is not repaired, the secondary failure will reoccur. For example, a plugged PCV passage can cause high crankcase pressure, resulting in leaking gaskets and seals. Replacing the gaskets and seals may stop the oil leak, but if the root cause (the PCV restriction) is not diagnosed and repaired, the oil leak will eventually return.

Scan Tool A test instrument that is used to read powertrain control system information.

Scan Tool Data Information from the computer that is displayed on the scan tool, including data stream, OTC's, freeze frame, systems monitors, and readiness monitors.

Secondary Air Injection A system that provides fresh air to the exhaust system under controlled conditions to reduce emissions; it can be either pulse or air pump type.

Sequential Multiport Fuel Injection (SFI) A fuel-injection system that uses one electronically pulsed fuel injector for each cylinder. The injectors are pulsed individually.

Speed Density System A fuel-injection system that calculates the amount of air drawn into the engine using engine RPM, air temperature, manifold vacuum, and volumetric efficiency, rather than measuring the mass or volume of air directly with an airflow meter.

Three-way Catalytic Converter (TWC) A catalytic converter system that reduces levels of HC, CO, and NOx.

Throttle Actuation Control (TAC) Used to describe any system that uses an electric motor to control the throttle plate. This system uses an accelerator pedal position (APP) sensor to monitor the accelerator pedal. That signal is used by the control module to open and close the throttle accordingly. This system does not use a throttle cable.

Trip A driving cycle that allows an OBD-II diagnostic test (monitor) to run.

Variable Valve Timing (VVT) System Refers to any system that advances, retards, or changes the duration and overlap of the camshaft.